Contemporary Science Teaching Approaches

Promoting Conceptual Understanding in Science

Contemporary Science Teaching Approaches

Promoting Conceptual Understanding in Science

edited by

Funda Ornek

*Bahrain Teachers College
at the University of Bahrain*

Issa M. Saleh

*Bahrain Teachers College
at the University of Bahrain*

INFORMATION AGE PUBLISHING, INC.
Charlotte, NC • www.infoagepub.com

Library of Congress Cataloging-in-Publication Data

Contemporary science teaching approaches : promoting conceptual understanding in science / edited by Funda Ornek, Issa M. Saleh.
    p. cm.
  Includes bibliographical references.
  ISBN 978-1-61735-608-7 (pbk.) – ISBN 978-1-61735-609-4 (hardcover) – ISBN 978-1-61735-610-0 (ebook)
  1. Science–Study and teaching (Secondary) 2. Inquiry-based learning.
I. Ornek, Funda. II. Saleh, Issa M.
  Q183.3.A1C67 2011
  507.1'273–dc23

                                                              2011035152

Printed in the United States of America

CONTENTS

Foreword vii

PART I

STRATEGIES IN TEACHING SCIENCE

1 Science Teaching Approaches to Promote Conceptual
Understanding in Science.. 3
Issa M. Saleh and Funda Ornek

2 Differentiating Science Pedagogy 9
Anila Asghar

3 The Metacognitive Science Teacher: A Statement for
Enhanced Teacher Cognition and Pedagogy.................................. 29
Gregory P. Thomas

4 Teaching About Climate Change: An Action Research Approach.... 55
John Wilkinson

5 Constructivist, Analogical, and Metacognitive Approaches
to Science Teaching and Learning 73
Samson M. Nashon and J. Douglas Adler

6 Concept Mapping and the Teaching of Science 115
Sadiah Baharom

7 Physics Modeling: An Approach Stimulates Students'
Conceptual Understanding of Scientific Concepts 137
Funda Ornek

PART II

USING COMPUTERS IN TEACHING SCIENCE

8 A Framework for the Integration of Technology into Science
Instruction 165
Yilmaz Saglam and Servet Demir

9 Real-Time Experiments and Images (RTEI) as Open Learning
Environments for Building Physics Knowledge 179
Giorgio Olimpo and Elena Sassi

10 Enhancing Asynchronous Learning in a Blended Learning
Environment 213
Mun Fie Tsoi

FOREWORD

For the last two decades of the 20th century, the developed world of mature public education systems contemplated the millennium, considering how best to deal with the profound and significant changes that were anticipated. Whether referring to the "global knowledge economy," the "innovation economy" or the more realistic and dynamic combination of both, there was increasing awareness that education, and the effectiveness of teaching and learning, was going to be key to preparing citizens for successful participation and engagement in whatever lay ahead. A consensus emerged that mathematics and science were core disciplines in the contemporary learning framework, and that science literacy was an essential attribute of the educated citizen of the new century, not just the province of elite engineers or academicians. There followed a renewed emphasis on the importance of the STEM disciplines for economic and social advancement, and a new focus on both the acquisition and application of scientific knowledge.

While the blueprint for a new balance of competencies, skills and abilities to inform curricula is still being articulated, with the critique came the recognition that science teaching was not necessarily leading to science learning as evidenced for many nations in the performance of their students in international tests such as TIMMS and PISA. Accordingly, a renewed and continuing interest in the profession of teaching, the preparation of teachers and the importance of the quality of teaching and learning has emerged; the overriding question is no longer why we teach mathematics and science, but rather how we teach them. This has become a global issue as developing systems seek to transform old pedagogies and practices into contemporary approaches to learning in order to create a teaching profession that is truly equipped for the tasks of the 21st century.

Contemporary Science Teaching Approaches, pages vii–viii
Copyright © 2012 by Information Age Publishing
All rights of reproduction in any form reserved.

Focus on science as an area meriting special attention in teacher preparation has evolved from the notion that a disciplinary background, whether a college degree or a content concentration, was the key and sometimes only entry standard. There is now a more comprehensive sense of the diverse set of skills and insights the effective science teacher needs to possess in order to meet the objectives of successful transfer of knowledge, cognitive skills and the ability to apply and problem solve in the real world. The realization that science can be taught outside of a laboratory, and that it can be joined to issues that engage students on a daily basis, is a part of the change that this century is embracing.

It is of note that this volume has been selected and prepared by two individuals who are part of the global process of transforming science teaching. As members of faculty at the Bahrain Teachers College, Funda Ornek and Issa Saleh are daily engaged in designing curricula, implementing new practices and measuring impacts that will profoundly alter the profession of teaching in the Kingdom of Bahrain and beyond. The depth and pace of the reform together with the ambitious goal of preparing truly reflective and contemporary teaching professionals provides a rich context in which to assess not only the impact of new pedagogies, but also the importance of learning cultures.

Ornek and Saleh, together with the contributors to this volume, are creating and testing new approaches to teaching science against the backdrop of a revised definition of learning outcomes just as they are using the full spectrum of instructional contexts to support new models: everything from IT through the living laboratory of the natural world, from headline news stories to the actions and interactions of everyday life are drawn into the mix. When joined to a base of understanding and reflection about learning, how it occurs, how it is supported and how it is assessed, their work constitutes a significant and practical contribution to creating new benchmarks of "best practice."

The contributors here offer to programme designers, policy makers and classroom teachers ideas drawn from the relationship of theory and practice to effective learning. They describe models that can be replicated, but that will also stimulate new thinking about everything from learning spaces through lesson planning; they offer indicators of how teachers can adapt and combine theoretical frameworks, practical examples and new perspectives as they meet the challenges of new expectations. Finally, the authors and the editors acknowledge and support the work of those many teachers who, through their constant search for new ways to foster effective learning, are true innovators and the essential catalysts in the task of human capital development.

—**Professor Kathryn Bindon**, PhD
Advisor to the President
University of Bahrain

PART I

STRATEGIES IN TEACHING SCIENCE

CHAPTER 1

SCIENCE TEACHING APPROACHES TO PROMOTE CONCEPTUAL UNDERSTANDING IN SCIENCE

Issa M. Saleh and Funda Ornek
Bahrain Teachers College, University of Bahrain

INTRODUCTION

Contemporary science teaching approaches focus on fostering students to construct new scientific knowledge as a process of inquiry rather than having them act as passive learners memorizing stated scientific facts. Although this perspective of teaching science is clearly emphasized in the National Research Council's National Science Education Standards (NRC, 1996), it is however challenging to achieve in the classroom. Science teaching approaches should enhance students' conceptual understanding of scientific concepts that can be later utilized by students in deeper recognition of the real world (Taylor, 2007). This book identifies and describes several different contemporary science teaching approaches, such as inquiry-based science teaching and constructivism, and presents recent applications of these approaches in promoting interest among students. It

Contemporary Science Teaching Approaches, pages 3–8
Copyright © 2012 by Information Age Publishing

promotes conceptual understanding of science concepts and conceptual change among them as well.

Chapter 2, "Differentiating Science Pedagogy" by Anila Asghar presents the challenges that students with learning disabilities face regarding science literacy, scientific reasoning, and hands-on and minds-on activities. Furthermore, the chapter enlightens some evidence-based practices that could be applied to promote inclusive science learning, for instance, advance organizers-graphic organizers, concept maps, jigsaw puzzles, cooperative scripting, and peer-instruction strategies.

Gregory P. Thomas, in his Chapter 3, "The Metacognitive Science Teacher: A Statement for Enhanced Teacher Cognition and Pedagogy," argues that metacognition is necessary to help enable "teachers to engage in practices that are consistent with achieving student learning outcomes." He further argues that, to some extent, classroom learning environments facilitate the development of metacognition by all students. In this chapter, he discuss the idea that in order for students to increase their understanding across science learning contexts and effectively learn science, they need to develop and enhance their adaptive metacognition. For that, the author recommends changing classrooms to be more "metacognitively oriented" for students to develop metacognition. He argues against the assumption that students will self-learn metacognition. The author indicates several reasons why very little has been written about teacher metacognition in science education. He gives three reasons, which are the research focus on improving "students' cognition, metacognition and learning outcomes," the fact that "teachers are inherently knowledgeable in relation to their own metacognition and cognition and also that of their students," and that "science teacher metacognition has been subsumed in other research such as that on teacher reflection, pedagogical content knowledge, and self-study." The author supports his argument about the importance of metacognition in science learning by referring to the work of Donovan and Bransford (2005), Georghiades (2004), White (1998), White and Frederiksen (1998), and White and Mitchell (1994).

Chapter 4, "Teaching About Climate Change: An Action Research Approach," by John Wilkinson addresses and reviews the impact of climate changes in society and how this change can be used to teach students content and science as a method of inquiry. The author argues that using climate changes as an exercise for students can help students to experience the complex nature of science. The author points out some of the specific skills that students might gain when introduced to such a topic. These skills are (a) students learn how to take random samples, (b) how to create a series of solutions, (c) observe the early life cycle of local freshwater plants, (d) use a control, and (e) organize and analyze data systematically. He acknowledges in his chapter the challenges that teachers might have to explain global climate changes to students. He points out the following challenges that teachers might face in order to implement it in classrooms: the science in this topic will be multi-disciplinary, computer models being beyond the scope of secondary science instruction, and the fact that data are constantly changing.

Chapter 5, "Constructivist, Analogical, and Metacognitive Approaches to Science Teaching and Learning," by Samson M. Nashon and J. Douglas Adler, provides constructivist, analogical, and metacognitive approaches for science teaching and learning. It discusses analogical strategies that include constructivist principles representing inquiry-based science pedagogy that promotes conceptual understating in science. Furthermore, this chapter presents the connections between analogical learning and conceptual change theory by applying analogical tools as teaching devices.

Sadiah Baharom, in her Chapter 6, "Concept Mapping and the Teaching of Science," examines concept mapping and its application in classrooms. She frames the chapter around Ausubel's assimilation-learning theory for meaningful learning and Novak's human constructivist view on concept mapping. She addresses the advantages and disadvantages of using concept mapping in classrooms. Some of the advantages, according to the author, are the fact that concept mapping promotes active learning: a concept map organizes information, visualization of knowledge, promotes higher-level thinking, and helps self-directed learning. The disadvantages that are expressed within the chapter are that concept mapping (a) takes a lot of instructional time, (b) can be laborious, and (c) is also a challenge that students might have in terms of the specific skills that are required before using concept mapping in classroom learning. She also looks at the six steps in constructing a concept map, which are (a) identifying concepts, (b) organizing concepts, (c) positioning concepts, (d) linking concepts, (e) revising concepts, and (f) finalizing. Finally, she sheds light on how concept maps can be quantitatively and qualitatively assessed, whicis an important component in teaching.

Chapter 7, "Physics Modeling: An Approach Stimulates Students' Conceptual Understanding of Scientific Concepts" by Funda Ornek, presents physics modeling, how it is applied in physics courses, how it can promote students' conceptual understanding of scientific concepts, and develop and challenge students' higher-order thinking in physics. Physics modeling is a contemporary teaching approach that provides authentic representation of contemporary physics. It also promotes students to construct their own knowledge through exploring physical phenomena by applying a few fundamental principles rather than given equations. This chapter also provides some modeling activities and relates some experiences from implementation of the physics modeling approach in teaching physics.

Chapter 8, "A Framework for the Integration of Technology into Science Instruction," by Yilmaz Saglam and Servet Demir, deals with the idea that technology integration plays a significant role in the development of professional programs. The purpose of the chapter is to provide science teachers with a framework for successful integration of technology into their instruction. According to the authors and the studies that they reviewed, scientific concepts are introduced to kids by animations, simulations, video clips, or models outside of schools. The authors also argue that these technology artifacts help in the understanding of science concepts by the students. The authors support the idea that technological artifacts facilitate students' understanding of science by looking at the work of Akpan and Andre (2000), Barnea and Dori (1999), Huppert, Lomask, and Lazarowitz (2002), Kelly and Jones (2005), and Sanger, Phelps, and Fienhold (2000). However, teachers in the classrooms are still resisting the use of technology in today's classrooms, according to Bell and Trundle (2008). The authors also review literature that points out that in order for teachers to be more effective in classrooms, they have to "possess technological pedagogical content knowledge (TPCK)." The authors also address the question, "What do teachers need to know for teaching with technology?" and "How can teachers develop TPCK?"

In Chapter 9, "Real-Time Experiments and Images (RTEI) as Open-Learning Environments for Building Physics Knowledge," Giorgio Olimpo and Elena Sassi argue the imporatance of experimental laboratory activities for the construction of physics knowledge. They introduce real-time experiments and images (RTEI), which is a type of labwork that helps to facilitate educational laboratory activities for students. The authors discuss the challenges of physics education and also outline the opportunities offered by RTEI. The condition in which RTEI can be utilized properly, according to the authors, would be under a student-centered approach. Also, they look at the different knowledge ranges that can be achieved by RTEI, such as "topical knowledge to networked knowledge to meta-knowledge." Finally, the authors give the following recommendations: "to provide teachers with a sound disciplinary knowledge," "to provide teachers with resources," and "to foster sharing and collaboration among teachers."

Chapter 10, "Enhancing Asynchronous Learning in a Blended Learning Environment" by Mun Fie Tsoi, provides an evidence-based research model, which is the TSOI Hybrid Learning Model (TSOI HLM), to enhance asynchronous learning in a blended learning environment. The TSOI HLM comprises four cognitive cyclical process phases based on constructivism. These phases are Translating, Sculpting, Operationalizing, and Integrating. Its most important feature is promoting active cognitive processing and addressing learning style. In this chapter, an authentic example from a science course that preservice science teachers have taken is presented by applying the HLM that guides the blended learning design involving asynchronous learning using Web 2.0. You will find in this chapter that the TSOI HLM has the functional potential capacity to give the instructors, teachers, and educators an alternative practice model for enhancing asynchronous learning in a blended learning environment. Students or learners will build on the various concrete experiences and learn how to construct new knowledge and integrate the knowledge with their existing knowledge in other contexts. Moreover, they will be active learners engaged in the various learning processes, including cooperative, collaborative and reflective practice.

In conclusion, the chapters in this book are the culmination of years of extensive research and development efforts in science education by distinguished science educators and researchers. The book attempts to develop a deeper knowledge of current best practice in science education and introduce contemporary pedagogical approaches in teaching science and science-related subjects. It also emphasizes science teaching through collaborative participation in a range of contexts and experiences. This book offers a powerful set of tools for teaching and learning science that can fully satisfy the needs of teachers, educators, and students in contemporary science teaching and learning.

REFERENCES

Akpan, J. P., & Andre, T. (2000). Using a computer simulation before dissection to help students learn anatomy. *Journal of Computers in Mathematics and Science Teaching, 19*(3), 297–313.

Barnea, N., & Dori, Y. J. (1999). High-school chemistry students' performance and gender differences in a computerized molecular modeling learning environment. *Journal of Science Education and Technology, 8*(4), 257–271.

Bell, R. L., & Trundle, K. C. (2008). The use of a computer simulation to promote scientific conceptions of moon phases. *Journal of Research in Science Teaching, 45*(3), 346–372.

Donovan, M. S., & Bransford, J. D. (2005). Introduction. In M. S. Donovan & J. D. Bransford (Eds.), *How students learn: History, mathematics and science in the classroom* (pp. 1–27). Washington DC: The National Academies Press.

Georghiades, P. (2004). From the general to the situated: Three decades of metacognition. *International Journal of Science Education, 26*(3), 365–383.

Huppert, J., Lomask, S. M., & Lazarowitz, R. (2002). Computer simulations in the high school: Students' cognitive stages, science process skills and academic achievement in microbiology. *International Journal of Science Education, 24*(8), 803–821.

Kelly, R., & Jones, L. (2005). *A qualitative study of how general chemistry students interpret features of molecular animations.* Paper presented at the National Meeting of the American Chemical Society, Washington, DC, August 28–September 1, 2005.

National Research Council. (1996). *National science education standards.* Washington, DC: National Academy Press.

Sanger, M. J., Phelps, A. J., & Fienhold, J. (2000). Using a computer animation to improve students' conceptual understanding of a can-crushing demonstration. *Journal of Chemical Education, 77*(11), 1517–1519.

Taylor, A. (2007). Learning science through creative activities. *School Science Review, 79*(286), 39–46.

White, B., & Frederiksen, J. (1998). Inquiry, modeling, and metacognition: Making science accessible to all students. *Cognition and Instruction, 16*(1), 3–118.

White, R. T. (1998). Decisions and problems in research on metacognition. In B. J. Fraser & K. G. Tobin (Eds.), *International handbook of science education* (pp. 1207–1213). Dordrecht, The Netherlands: Kluwer.

White, R. T., & Mitchell, I. J. (1994). Metacognition and the quality of learning. *Studies in Science Education, 23,* 21–37.

CHAPTER 2

DIFFERENTIATING SCIENCE PEDAGOGY

Anila Asghar
McGill University

STUDENTS' INTUITIVE IDEAS ABOUT THE NATURAL WORLD

A pervasive assumption that has influenced the beliefs and practice of many, if not all, science educators, is that the formal science education in school is the main source of scientific ideas and models about the natural world. But as Driver, Squires, and Wood-Robinson (1994) state, "learning about the world does not take place in a social vacuum" (p. 3). Science educators must recognize that students' interactions and experiences with the natural world shape their ideas in significant ways. Often, science teachers exacerbate the situation by viewing each student as a *tabula rasa* to fill with principles and theories, or they presume perfect prior knowledge with which to build more-complex concepts upon. The possibility of students' intuitive ideas about natural phenomena is rarely acknowledged. Teachers must realize that children "don't just passively receive information," but instead "operate on it and transform it," (Baker & Piburn, 1997, p. 31) based on global and personal experiences. Driver (1985) contends that children's perceptual experiences influence their conceptual frameworks around sci-

Contemporary Science Teaching Approaches, pages 9–28
Copyright © 2012 by Information Age Publishing
All rights of reproduction in any form reserved.

entific models. Sadler and colleagues (Schneps & Sadler, 1997) conducted a thought-provoking study with school children and university graduates to elucidate how young people, from elementary school through college, perceive the relationship between scientific principles and the world around them. The interviews are captured in a video titled "Minds of Our Own— Lessons From Thin Air." The video shows a thread linking the *same* general misconceptions about basic science concepts, such as photosynthesis, light, states of matter, and seasons, from elementary students to graduating science and engineering university students. Even after years of top-notch schooling, these graduates' basic ideas about nature resonated with the ones held by elementary and middle graders.

The most important lesson for science educators, curriculum developers, and researchers is that many students do not develop a clear and robust understanding of the fundamental concepts in science. This challenges the assumption that science instruction can fill children's minds with science models that they will carry in their heads all their lives. Research in science education (Asghar & Libarkin, 2010; Fuson, 1988; Harlen, 1987; McCloskey, 1983; Smith, Maclin, Grosslight, & Davis, 1997), on the contrary, shows that students' experiences with natural phenomena and their everyday language heavily influence their understanding of science conceptual frameworks. According to McCloskey (1983), people develop remarkably well-articulated alternative conceptions on the basis of their everyday experience, and they are deep-seated and persist even after exposure to formal science instruction. When scientific models are presented to them, they interact with their preexisting notions in interesting ways. Their intuitive skills are integrated with the new information, often producing localized and incoherent frameworks, which they apply in an inconsistent manner to explain and solve problems. The *abstract and formal* language of science could exist alongside children's intuitive frameworks.

Scholars agree that children's ideas about the world "differ in conceptual content from those of scientists" (Brewer & Samarapungavan, 1991; Driver, 1981; Driver et al., 1994; Harris, 1994; McCloskey & Kargon, 1988; Thagard, 1989; Vosniadou, 2007; Vosniadou & Brewer, 1992; Wiser, 1988). Nevertheless, there is a continuing debate over the nature of children's ideas. Some researchers argue that children's intuitive knowledge is fragmented, unstable, and consists of weakly connected systems of ideas (Claxton, 1993; diSessa, 1983, 1993; Vosniadou, 1994). Other researchers contend that children's novice theories or "explanatory frameworks" exhibit some properties that are similar to scientific theories (Brewer & Samarapungavan, 1991; Nakhleh & Samarapungavan, 1999; Samarapungavan & Wiers, 1997; Wiser, 1988) that children may use to make predictions, develop cause-and-effect explanations, and apprehend and solve scientific problems (Samarapungavan & Wiers, 1997). These intuitive frameworks interact with the new concepts that students learn in science classes in interesting ways, often making it difficult for children to develop expert knowledge.

Research on K–12 and college students' understanding of scientific models suggests that alternative frameworks about fundamental physical, geological, and biological concepts may persist despite years of exposure to science instruction in formal education settings (Asghar, Libarkin, & Crockett, 2001; Driver, 1985; Driver & Russell, 1982; Osborne, 1980; Perkins, 1992). Children's underlying preconceptions are resistant to change, and they influence their learning of new scientific ideas in important ways (Asghar & Libarkin, 2010; Libarkin & Asghar, 2002). Asghar and Libarkin, for instance, found that college students carried several underlying misconceptions about the concepts of force, velocity, acceleration, and gravity. Moreover, their ideas about gravity were strikingly similar to elementary students' intuitive conceptions of gravity.

CHILDREN'S ALTERNATIVE CONCEPTIONS OF CONSERVATION OF MASS

Findings from a study focusing on children's alternative frameworks related to the conservation of mass are discussed herein to illustrate the ways in which they interfere with their developing understandings of this concept. Scientists view matter as composed of indestructible particles (atoms) that are unique to and characteristic of each element. The total number of atoms in a system is conserved during every physical or chemical transformation. By accounting for these particles, chemists make accurate predictions about the physical world. For example, scientists can calculate the optimal ratios of chemicals needed for certain reactions so that no components remain uncombined. Research shows that a considerable proportion of students hold a fairly stable model of matter in which matter is perceived as "continuous and static" (Children think that a substance is composed of matter that can be infinitely divided into small pieces of the same material—a non-particulate view. Water and steam are viewed as having entirely different compositions). This model influences the learning of the "abstract" particle model of matter in school (Driver, 1985, p. 36). Researchers have found that children (9–13 years) possess inconsistent frameworks about matter across a range of substances from "continuous solids to particulate solids" (Nakhleh & Samarpungavan, 1999; Driver, 1985). Piaget and Inhelder (1974) state that children's perceptual and concrete experiences influence their reasoning. Children tend to make sense of a substance by observing their macroscopic properties, and if a substance disappears during a physical or chemical change, it is no longer the same.

Most 7- to 13-year-old children hold a continuous view of matter (Briggs & Holding, 1986). Qualitative studies reveal the range of children's non-particulate ideas. Children think, for instance, that a solid is either lost completely or loses weight while dissolving and melting (Driver et al., 1985; Stavy & Stachel, 1984). Children possess interesting preconceptions around combustion. They tend to think that metals would become lighter after burning because "certain things" are "burnt away" (Driver, 1985, p. 38).

Children also possess diverse ideas about the dissolving of sugar and other substances in water at various stages of development. Children up to age 8 describe the dissolving of sugar as something that "just goes," "disappears," "melts away," "dissolves away," or "turns into water" ((Driver, Squires, Rushworth, & Wood-Robinson, 1994). Older students (12–16 years) think that sugar "goes into tiny little bits" as it dissolves or "mixes with water molecules" (Abraham & Williamson, 1994; Bar, 1986; Briggs & Holding, 1986; Driver et al, 1994; Nussbaum, 1985). Another frequently expressed idea is that a solid loses weight when it changes into liquid state (Stavy & Stachel, 1984). Some children think that sugar is "up in the water" and is not "pressing down" on the bottom of the vessel (Abraham & Williamson, 1994). Some students think that the sugar may still be present in water after dissolving but is "lighter." A plausible explanation for this reasoning is that after dissolving in water, sugar is not a solid, its mass therefore cannot act on a surface in the same way (Driver, 1985).

For some children, a solution is a "single substance" rather than a mixture of different substances. The non-particulate nature of a solution is more prevalent among 10–12 year olds (Briggs & Holding, 1986). When students' (age 9–14 years) "quantitative reasoning" was explored through a survey, about two thirds of them predicted that the mass of the solution will be less than the individual components (Driver & Russell, 1982). Surprisingly, students aged 14 to 17 years also demonstrated a continuous non-particulate view and were not able to conserve the mass of substances while explaining physical and chemical reactions (Ben-Zvi, Eylon, & Silberstein, 1987). Children can integrate the particulate view into their existing frameworks about perceptible properties of matter in an interesting manner. Although most secondary school students interpret chemical changes using the particle model of matter, some tend to assume that the particles possess the same macroscopic properties of that substance (Ben-Zvi et al., 1987; Driver et al., 1994). Their particulate framework thus is not consistent with the scientific model about conservation of mass. They might assume that particles shrink or melt during physical and chemical changes (Ben-Zvi et al., 1987). Without a particulate view of matter, children usually are not able to conserve the mass of substances during physical and chemical transformations of matter. This model affects the learning of the chemical principles in school (Nussbaum, 1985). If they think, for instance, that gases have no weight, they are not likely to conserve the overall mass of the gases involved in a chemical reaction (Driver, 1985).

Research suggests that developing a particulate view of matter would help students to understand, predict, and explain changes in the appearance of substances during physical and chemical changes as the reorganization of discrete particles (Driver, 1985; Nussbaum, 1985). In studying student ideas about conservation, Piaget (1973) maintained that children who initially constructed a particulate view of matter would more easily develop the idea of conservation of mass. Thus, changes in physical appearance or creation of new substances during a chemical reaction could be understood in terms of rearrangement of indestructible particles. An understanding of the particulate nature of matter is the foundation upon which the understanding of chemical and physical change rests (Driver, 1985). Therefore, in modern chemistry, the idea that matter is particulate and continuous is the fundamental model that explains any kind of transformations in matter (Driver, 1985; Nussbaum, 1985).

A study was conducted with middle school students to look at how their understanding of conservation developed while engaging with two different types of chemistry curricula. One curriculum focused on the traditional methods of instruction while the other curriculum employed concrete models present in the particulate nature of matter. The traditional chemistry curriculum focused on basic chemistry concepts, such as physical change, chemical reaction, and conservation of mass. The instruction constituted lectures, demonstrations, reading, worksheets, and questions included in the science textbook. The study also investigated the developmental trajectories of students' ideas as they participated in an inquiry-based curriculum using physical models to demonstrate a concrete view of matter as composed of particles. This curriculum also focuses on basic chemistry concepts related to physical and chemical changes and is designed to help students gain an understanding of the conservation of mass. Students are engaged in hands-on experiences, such as mixing materials to observe physical and chemical changes, and use a concrete physical model to understand and explain the changes at the molecular level. This model helps them to understand how chemical change works. The most important concept that they learn by using this physical model is that the total number of particles involved in the reaction is conserved before and after the chemical change (i.e., the total mass of the system is conserved). This concrete model is designed to help students to learn about the scientific concept of conservation of mass not as an abstract scientific principle, but as a concrete model that they could apply to explain and predict the results of experiments using the conservation of mass principle. Using this model, they can test different ratios of the reactants and predict the amount of the products in simple chemical reactions, such as the reaction between baking soda and vinegar. Students interact in small inquiry groups to conduct activities and engage in discussions around their experimental design and results.

This study was conducted in four middle-grade classrooms from two public schools in Massachusetts. Each teacher in the study taught two eighth-grade science classes and used the inquiry/model curriculum in one class and the traditional curriculum in the other. The four classes in both schools were heterogeneous in terms of ethnicity (e.g., Caucasian, African American, Latino, Asian, Haitian, mixed, etc.). A total of 73 students in four classes participated in the study, of which 38 were girls and 35 boys. The average number of students in each class was 25 at one school and 11 at the other school.

Data were collected through (a) administering a conceptual questionnaire and (b) qualitative interviews with 16 students (four from each class) before and after the curriculum. The questionnaire assesses children's understanding related to conservation of mass during physical and chemical changes (some conservation tasks are described later in the concept maps section). Semi-structured interviews lasting about 45–60 minutes were conducted to explore their ideas and emotions in-depth. Four students from each class were selected for this purpose. A variety of analytic strategies were employed to carry out a detailed analysis of the data in relation to each research question. Students' responses to the multiple-choice items in the concept questionnaire were scored. If they conserved they received a score of one, if they did not conserve the mass of substances, they were assigned a score of 0. The interview transcripts from the 16 selected students and their open-ended responses on the pre- and post-questionnaire were analyzed using qualitative strategies. Open coding and categorizing techniques (Strauss & Corbin, 1998) were used to code and organize data for the cognitive dimension in the individual responses to the open-ended questions in the concept questionnaire and interviews. Codes were grouped into themes (e.g., instances where matter is conserved) and themes were batched into categories. Field notes, analytic memos, and visual displays (Maxwell, 2005; Miles & Huberman, 1994) helped to track and compare participants' developing understanding of conservation within and across cases.

Middle school students participating in this study expressed several interesting intuitive conceptions about basic chemistry ideas. For example, most children said that air has no weight. This misconception interacted with their developing understanding of the conservation of mass in interesting and surprising ways. For instance, while predicting the mass of an open can of cola, many children thought that there would be no change in the total mass of the system because air is weightless. This misconception influenced their thinking in relation to other conservation tasks as well. While considering the mass of a sealed jar that contained a burning candle, most children thought that the total mass of the system would be less after the combustion of the candle. This prediction was grounded in the misconception that the candle wax will melt and convert into a weightless gas resulting in a decrease in the total mass of the jar. Conversely, some children thought that pressure has mass and therefore predicted that the total mass of the jar would increase due to the increased mass of the smoke after the wax turned into smoke. Another frequently expressed misconception was that mass equates density. Some students predicted a decrease in the mass of the open cola can because they thought it would lose its density after losing air or soda. Some students, on the other hand, thought that the density, and thus the mass, of the open cola can would not change as atmospheric air would enter the open cola.

Another interesting finding was that students used disparate frameworks in relation to the same conservation problem and made conflicting predictions about the mass of the system. Hence, while predicting the mass of a closed jar after the combustion of a candle, some students considered only one component of the system while making their predictions. They thought that the mass of the jar would decrease because the size of the candle would be shorter after burning. Interestingly, they used different reasoning while considering the mass of a sealed test chamber after the combustion of candles. Some students said that the total mass of the system would increase because of the smoke produced from combustion.

The findings of this study suggest that the students developed and applied localized thinking structures about conservation of mass. The scientific model of mass conservation was presented to them in the inquiry-based curriculum through the concrete particulate model and in the traditional curriculum through the abstract atomic model. Nevertheless, many students from both curriculum groups did not develop a global understanding of the conservation principle. Their preexisting skills about various phenomena, such as combustion and melting, interacted with their newly acquired ideas about the conservation of mass producing localized frameworks related to specific problems on the concept test. This means that they conserved the mass of substances on some problems while simultaneously demonstrating a non-conserving view in relation to other problems. They did not develop a coherent and generalized understanding of the conservation of matter.

The preconceptions held by these students resonated with the findings of the relevant literature (Driver, et al., 1994; Smith, 1988). The most prevalent misconceptions expressed by the students were that (a) air is weightless, (b) hot air is lighter, (c) density is equated with mass, (d) substances gain or lose mass after dissolving, (e) solids lose mass after melting, (f) the amount of melted water is greater than the amount of frozen ice, and (g) air pressure contributes to the mass of a system. Students mostly tended to focus on a single component of the system rather than considering the entire system. The existing literature on children's ideas corroborates these findings (Bar, 1986; Briggs & Holding, 1986; Carey, 1985; Driver, 1985; Driver & Russell, 1982; Smith et al., 1997; Stavy & Stachel, 1984). Some of these misconceptions persistently showed up in students' responses before and even after participating in both types of curricula.

INSTRUCTIONAL STRATEGIES FOR PROMOTING CONCEPTUAL CHANGE IN INCLUSIVE SETTINGS

Changing children's or adult's intuitive conceptions about the natural world is not an easy feat to accomplish. Different models of conceptual change have been proposed by cognitive-science experts. The conceptual change model is based on creating a dissonance between learners' existing conceptions and the new scientific ideas. According to this model, a meaningful conceptual change in children's scientific ideas requires that they must be dissatisfied with their existing ideas and appreciate the plausibility and intelligibility of the scientific conception (Hewson, 1981, 1982; Strike & Posner, 1985). Conceptual learning is generally understood as a process (Carey, 1985; Demastes, Good, & Peebles, 1996; Strike & Posner, 1992) that encompasses "conceptual assimilation" (integration of a new conception with learner's existing ideas) and "conceptual accommodation" (replacement of existing ideas by the new conception). A fundamental prerequisite for any conceptual change, nevertheless, is that students become aware of their existing ideas so that they can think about the ways in which they are using them to understand and approach any problem-solving tasks while learning science. Helping students to uncover and share their preconceptions and examine how their ideas change or develop further by making connections with other relevant ideas is thus an important goal of science instruction (Stepans, 2003).

Concept Maps and Graphic Organizers: Science teachers can employ different tools, such as concept maps, graphic organizers, visual displays, and drawings to help students to articulate and examine their preconceptions and developing scientific conceptions. These tools and strategies can be used effectively in regular as well as inclusive classroom settings as *all* children tend to carry intuitive preconceptions and need to be supported while learning scientific concepts. Nevertheless, research suggests that students with learning disabilities may have preconceptions that are "several years below those of their age peers" (Scruggs, Mastropieri, & Okolo, 2008, p. 3). Furthermore, children with mild and high-incidence disabilities also seem to face difficulty with developing critical thinking and independent reasoning associated with problem-solving tasks in science. Concept maps and graphic organizers can be particularly helpful in developing specific accommodations for students with special needs to facilitate their conceptual understanding of scientific models (Anderson-Inman, Ditson, & Ditson, 1998; McLeskey, Rosenberg, & Westling, 2010). Graphic organizers are tools to clarify and organize prior ideas so that new ideas can be integrated with relevant existing concepts (Ausubul, 1963).

Concept maps and graphic organizers have been shown to be effective in developing students' conceptual development in different content domains. Concept maps are valuable in supporting students to comprehend and represent the relationships among various concepts (Guastello, 2002; Seamen, 1990; Yates & Yates, 1990). Concept or cognitive maps serve as useful tools to visually represent the relationships among different components of a scientific model. They are mainly valuable in helping students to express their mental constructs and meaning-making in a graphic form. Connections among different concepts are displayed through lines or arrows to create a network of closely connected ideas (Novak, 1990, 1991, 1993, 1998). Different formats and designs can be used for constructing concept maps (Figure 2.1).

Concept maps can be used as assessment and pedagogic tools in science instruction because they make it possible to cover and evaluate the text more efficiently (Roberts & Joiner, 2007). Gerstner and Bogner (2009) suggest that concept maps may serve as effective tools for assessing students' learning achievement in science when used in combination with other assessment tools such as multiple-choice tests. Concept maps are also useful in assessing and displaying children's naïve preconceptions before instruction. The following concept map illustrates a common misconception among children about the mass of sugar after dissolving in water. These ideas were expressed by the middle school students in the study described above while predicting the mass of a sugar cube after it dissolved in the water.

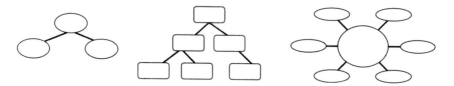

Figure 2.1 Concept map designs.

Task: Mass of a Sugar Cube After Dissolving in Water

Sue balances a cup of water with a sugar cube outside with a cup of water with a sugar cube in it. After the sugar cube dissolves, what will happen to the pan on the right?

Preconception

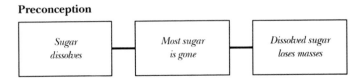

The teacher can also use a concept map as an instructional tool to exhibit the conservation of mass concept after presenting relevant hands-on activities demonstrating that the total mass of a system is conserved after sugar dissolves in water. Participation in hands-on activities and experiments is vitally important to develop students' thinking, writing, and communicating skills in science (McCarthy, 2005; Webb, 2010).

Conservation of Mass Concept:

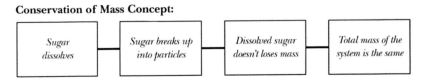

Similarly, children's misconceptions about the mass of air and the conservation of mass concepts in the context of the open cola can and combustion tasks described earlier in the discussion of the findings from the study focusing on the conversation of mass can be displayed as follows:

Task: Mass of an Open Can of Cola After it Goes Flat

Fatima balances two cans of cola, one closed and one just opened. After the open can of cola goes flat, what will happen to the pan on the right?

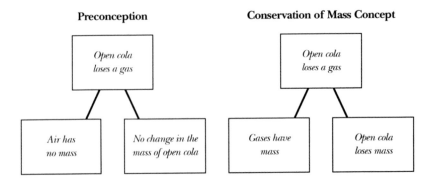

Providing opportunities to students to revisit their misconceptions about conservation of mass alongside examining the scientific model of conservation might help them to confront their alternative conceptions and see the underlying representations that contributed to their misconceptions. They can also compare the differences between their misconceptions and the scientific conceptions. Teachers in K–12 and university settings can employ concept maps to challenge and address students' intuitive conceptions and facilitate a deeper understanding of the accepted scientific models.

Task: Combustion of a Candle in a Sealed Jar

Nico balances two sealed containers, one with an unlit birthday candle and one with a lit candle. After one candle burns halfway down, what will happen to the pan on the right?

Preconception

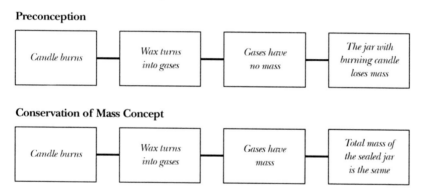

Conservation of Mass Concept

Horton, Lovitt, and Bergerud (1990) looked at the effects of graphic organizers as a pedagogical tool in a study with secondary students enrolled in content courses. Their findings suggest that graphic organizers led to higher student performance than self-study groups. Furthermore, graphic organizers were particularly useful in supporting students with learning disabilities in physical science classes (Horton et al., 1990). Similarly, Guastello (2002) reported the effectiveness of using concept maps in science instruction. Students in the intervention group used concept maps to learn and express their understanding of science concepts and performed better than the control group that was exposed to the traditional teacher-directed method. Concept mapping helped in improving low-achieving middle school students' conceptual comprehension as they actively participated in assimilating and constructing new scientific ideas.

Concept mapping has also been shown to be an effective strategy in learning, retaining, and recalling knowledge related to different domains, such as biology, chemistry, and physics (*Keraro, Wachanga, & Orora*, 2007; Lindstrom & Sharma, 2009; Markow & Lonning, 1998). Lindstrom and Sharma looked at the effects of using "link maps" in physics instruction in a study with first-year college students. Link maps—a physics-specific aid—were used to illustrate the interconnectedness of fundamental ideas in physics. The link maps contain fewer concepts than conventional concept maps because the purpose is to foster students' conceptual understanding of the key concepts rather than encouraging rote learning. Weekly supplementary enrichment tutorials—"map meetings"—were offered to present the materials visually using link maps. Link maps were found to be useful in advancing student understanding of abstract physics concepts and problem-solving skills, according to Lindstrom and Sharma. This study suggests that map meetings helped in creating a constructive learning environment for physics novices. Conversely, Woodward (1994) contends that generic techniques used for modifying science texts, such as direct instruction, mnemonics, and graphic organizers, may not be sufficient to deal with the "increasing complexity" of secondary science materials. Woodward argues that context-rich problem solving might be more effective in developing a rich understanding of science in secondary students with learning disabilities.

Students with learning disabilities often struggle with organizing abstract ideas and facts into coherent frameworks, and many might approach science content as a "set of isolated facts." Smith and Okolo (2010) suggest that technology-based tools could be embedded in effective research-based practices, such as "graphic organizers," "strategic and procedural support for writing," and "explicit instruction" to create better learning opportunities for students with learning disabilities. For example, Web-based writing programs provide a variety of tools to support the struggling writer, such as graphic organizers, thinking sheets, checklists, and feedback on emerging ideas; and help with organizing, revising, and editing written assignments (Smith & Okolo, 2010). Scruggs, Mastropieri, Berkeley, and Graetz (2010) conducted a comprehensive meta-analysis of the literature to assess the effectiveness of different interventions designed for special needs students. Their findings indicate that spatial and graphic organizers, mnemonic devices, and computer-assisted instruction had a significant impact on student learning in the areas of science, social studies, and English (Scruggs et al., 2010). Online tools, such as WebQuests are being increasingly used by teachers to engage children in guided learning activities. Skyler, Higging, and Boone (2007) propose that "graphic organizers, hypertext study guides, outlines, vocabulary definitions, annotated lists of Web sites, and templates for compiling information" could be used to adapt and simplify complex WebQuest tasks for students with learning disabilities.

Chin and colleagues (2010) studied the effects of online interactive concept maps—"Teachable Agents" (TA)—on students' science learning. TA is an instructional tool that engages children in teaching their agent by creating concept maps. Their agent can answer questions using the concept maps, thus enabling children to get feedback on their understanding. Students learn by teaching a computer character. The students create the concept map that is the character's "brain," and they receive feedback based on how well their computerized pupil can answer questions" (Chin et al., 2010, p. 651). The study shows that TAs enabled students to learn new concepts from their regular science lessons. Furthermore, TAs improved student understanding of causal relations among different concepts.

Concept maps can also help in vocabulary development; they can be used for creating vocabulary banks to record the scientific terms and their meanings. Students can also collect and add pictures that illustrate the concepts. Developing vocabulary webs will help them to represent how they connect different words and concepts (Moore & Readence, 1984; Sabornie & deBettencourt, 2009). Studies suggest that vocabulary books and banks can increase reading comprehension and vocabulary recall if used as a regular pedagogical strategy (Beck & McKeown, 1991; Beck, Perfetti, & McKeown, 1982).

Strategies to Support Understanding of Scientific Texts: Comprehension of scientific texts is a persistent challenge that students face in regular as well as special education settings. According to Snow (2010), scientific language is "concise, precise, and authoritative." Sophisticated and complex scientific and technical terms can "disrupt reading comprehension and block learning" (p. 405). Science teachers need to develop and provide specific support systems to help students to understand the scientific language and become independent learners of science. Understanding the meaning of the scientific words and concepts and how they are connected to each other is essential to meaningful science learning. Many middle and secondary school students and even college students struggle with academic scientific jargon and language. Snow suggests that literacy and science education experts need to collaborate to develop strategies to assist students to "convert their word-reading skills into comprehension" while reading scientific texts. The ability to read, write, and communicate scientific ideas constitutes important features of scientific literacy (Webb, 2010). Students must be able to access the formal and specialized language of science as they engage in hands-on and minds-on science. Proficiency in scientific discourse is essential to pursuing scientific investigations. Furthermore, students must be encouraged and supported in K–12 education to ask questions and pursue inquiry-based investigations around their questions. However, providing opportunities for recording and writing about their results is equally important in science education (Webb, 2010).

Research suggests that students with disabilities lag behind their peers in middle and secondary science (Mastropieri et al., 2006). Moreover, students with disabilities struggle with scientific terms and formulae even in effective learning environments (Mastropieri et al., 2006; Scruggs, Mastropieri, Bakken, & Brigham, 1993). Differentiating science instruction by using research-based pedagogical strategies will help educators in creating different kinds and levels of supports to help *all* learners to successfully engage in science learning activities (Lawrence-Brown, 2004; Rosenberg, Westling, & McLeskey, 2008; Waldron & McLeskey, 2001).

Several studies demonstrate that strategies related to specific tasks, such as reading and concept acquisition, can facilitate student engagement and learning in specialized content domains like mathematics and science (Borkowski, Weyhing, & Carr, 1988; Hughes & Schumaker, 1991; Mastropieri, Scruggs, & Shiah, 1991). Summary skills strategy has been developed to actively engage students experiencing learning difficulties in science (Nelson, Smith, & Dodd, 1992). This learning strategy has been useful in assisting students with disabilities. Specifically, it helps students to organize, understand, and recall important ideas and information in science texts. The strategy involves identifying and summarizing the main ideas in a text in students' own words; unnecessary details are avoided so that students can focus on the key concepts presented in a scientific text. Nelson and colleagues studied the effects of summary skills strategy on elementary special education students' understanding of science text in an urban setting. They found this strategy to be highly effective in improving students' understanding as well as their writing skills.

Bakken, Mastropieri, and Scruggs (1997) examined the effects of text-structure-based reading strategies on middle school students' comprehension of science and social studies texts. Their findings suggest that this strategy is useful in helping students with disabilities who face difficulties in understanding science vocabulary, concepts, and the supporting evidence for scientific arguments. This strategy involves identifying the main ideas and their supporting evidence in a text. Students locate the main idea and supporting evidence after reading a passage and summarize this information in their own words. The study shows that text-structure-based reading strategies significantly improved students' recall of main and incidental information on immediate, delayed, and transfer assessments as compared to the traditional instructional methods in science. The authors maintain that middle school students with learning disabilities can "learn, apply, and transfer text-structure-based strategies" (Bakken et al., 1997).

The repeated-reading strategy is also beneficial for students with mild learning disabilities. Research suggests that repeated reading actively engages students in their learning and improves their reading fluency skills. This strategy involves reading a particular text multiple times. Students can also discuss the main ideas during this process. Repeated reading can serve as a useful strategy to enhance students' understanding in specialized content areas, such as science and social studies (Sabornie & deBettencourt, 2009; Sindelar, Monda, & O'Shea, 1990).

Lovitt and Horton (1994) argue for modifying content textbooks for youth with mild disabilities in inclusive classrooms. Students with learning disabilities often experience difficulties with reading and textbooks, and assimilating new information with their existing mental schemas. They also recommend that general-education teachers should be involved in the process of adapting science textbooks. Visual tools, such as graphic organizers are useful in making difficult and poorly organized passages more accessible to students. Textbooks in biology, physical and environmental science, and health are laced with sections that can be neatly organized by graphic organizers of one type or another (p. 115).

Peer-assisted learning (PAL) or peer tutoring strategies have been shown to be successful in mathematics and science learning. Mastopieri (2006) reports that peer tutoring can be employed effectively to differentiate hands-on activities to facilitate science-content learning for students with mild disabilities in inclusive middle-grade classrooms. PAL involves collaborative instruction in pair groups. Students needing academic assistance (tutees) are paired with other students who are proficient in a certain concept or skill (tutors) who use instruction and feedback related to the task to help the tutees. The tutor and tutee switch their roles over time in this process also (Calhoon & Fuchs, 2003; Sabornie & deBettencourt, 2009).

Mnemonic or memory devices have been shown to be particularly useful in different science domains like geology, biology, paleontology, and chemistry. They are highly effective in retaining and recalling information (Mastropieri, Scruggs, & Levin, 1987; Mastropieri, Scruggs, & Graetz, 2005; Sabornie & deBettencourt, 2009; Scruggs et al., 2008). For example, a useful mnemonic for remembering the names and order of the planets in the solar system is: *My Very Efficient Mother Just Served Us Nuts.* Similarly, the mnemonic for remembering the taxonomic classification scheme in biology (Kingdom, phylum, class, order, family, genus, and species) could be *King Peter can order fish, greens, salsa.*

Science educators can employ and adapt these evidence-based strategies to enhance all students' scientific literacy, conceptual development, and reasoning skills. Further research is needed to study the effects of these differentiation strategies to learn about the ways in which specific accommodations can help students with learning disabilities in science classrooms.

REFERENCES

Abraham, M., & Williamson, V. (1994). A cross-age study of the understanding of five chemistry concepts. *Journal of Research in Science Teaching, 3,* 147–165.

Anderson-Inman, L., Ditson, L. A., & Ditson, M. T. (1998). Computer-based concept mapping: Promoting meaningful learning in science for students with disabilities. *Information Technology & Disabilities, 5*(1–2).

Asghar, A., & Libarkin, J. (2010). Gravity, magnetism, and "down": College students' conceptions of gravity. *The Science Education, 19*(1), 42–55.

Asghar, A., Libarkin, J., & Crockett, C. (2001). *Invisible misconceptions: Students' understanding of ultraviolet and infrared radiation.* Presented at the GSA Annual Meeting & Exposition Program, the Geological Society of America. Boston, MA. November 5–8, 2001.

Ausubel, D. P. (1963). *The psychology of meaningful verbal learning.* New York: Grune & Stratton.

Baker, D. R., & Piburn, M. D. (1997). *Constructing science in middle and secondary school classrooms.* Boston: Allyn and Bacon.

Bakken J. P., Mastropieri, M. A., & Scruggs, T. E. (1997). Reading comprehension of expository science material and students with learning disabilities: A comparison of strategies. *The Journal of Special Education, 31*(3), 300–324.

Bar, V. (1986). *The development of the conception of evaporation.* Jerusalem: The Hebrew University of Jerusalem.

Beck, I. L., & McKeown, M. G. (1991). Conditions of vocabulary acquisition. In R. Barr, M. Kamil, P. Mosenthal, & P. D. Pearson (Eds.), *Handbook of reading research* (Vol. 2, pp. 789–814). White Plains, NY: Longman.

Beck, I. L., Perfetti, C. A., & McKeown, M. G. (1982). Effects of long-term vocabulary instruction on lexical access and reading comprehension. *Journal of Educational Psychology, 74*(4), 506–521.

Ben-Zvi, R., Eylon, B., & Silberstein, J. (1987). Students' visualization of chemical reactions. *Education in Chemistry, 24,* 117–120.

Borkowski, J. G., Weyhing, R. S., & Carr, M. (1988). Effects of attributional retraining on strategy-based reading comprehension in learning-disabled students. *Journal of Educational Psychology, 80,* 46–53.

Brewer, W. F., & Samarapungavan, A. (1991). Children's theories versus scientific theories: Differences in reasoning or differences in knowledge? In R. R. Hoffman & D. S. Palermo (Eds.), *Cognition and the symbolic processes: Applied and ecological perspectives* (Vol. 3, pp. 209–232). Hillsdale, NJ: Erlbaum.

Briggs, H., & Holding, B. (1986). *Aspects of secondary students' understanding of elementary ideas in chemistry.* Leeds, UK: University of Leeds, Center for Studies in Science and Mathematics Education.

Calhoon, M. B., & Fuchs, L. S. (2003). The effects of peer-assisted learning strategies and curriculum-based measurement on the mathematics performance of secondary students with disabilities. *Remedial and Special Education, 24,* 235–245.

Carey, S. (1985). *Conceptual change in childhood.* Cambridge, MA: Bradford Books/MIT Press.

Chin, D., Dohmen, I., Cheng, B., Oppezzo, M., Chase, C., & Schwartz, D. (2010). Preparing students for future learning with Teachable Agents. *Educational Technology Research & Development, 58*(6), 649–669.

Claxton, G. (1993). Mini theories: A preliminary model for learning science. In P. J. Black & A. M. Lucas (Eds.), *Children's informal ideas in science* (pp. 45–61). London: Routledge.

Demastes, S. S., Good, R. G., & Peebles, P. (1996). Patterns of conceptual change in evolution. *Journal of Research in Science Teaching, 33*(4), 407–431.

diSessa, A. A. (1983). Phenomenology and the evolution of intuition. In D. Gentner & A. L. Stevens (Eds.), *Mental models* (pp. 15–33). Hillsdale, NJ: Lawrence Erlbanm Associates.

diSessa, A. A. (1993). Toward an epistemology of physics. *Cognition and Instruction, 10* (2–3), 105–225.

Driver, R. (1981). Pupils' alternative frameworks in science. *European Journal of Science Education, 3,* 93–101.

Driver, R. (1985). Beyond appearances: The conservation of matter under physical and chemical transformations. In R. Driver, E. Guesne, & A. Tiberghien (Eds.), *Children's ideas in science* (pp. 145–169). Milton Keynes, UK: Open University Press.

Driver, R., Guesne, E., & Tiberghien, A. (1985). *Children's ideas in science.* Milton Keynes: Open University Press UK.

Driver, R., & Russell, T. (1982). *An investigation of the ideas of heat, temperature, and change of state of children aged between eight and fourteen years.* Leeds, UK: University of Leeds, Center for Studies in Science and Mathematics Education.

Driver, R., Squires, A., Rushworth, P., & Wood-Robinson, V. (*1994*). *Making sense of secondary science: Research into children's ideas.* New York: Routledge.

Gerstner, S., & Bogner, F. (2009). Concept map structure, gender and teaching methods: An investigation of students' science learning. *Educational Research, 51*(4), 425–438.

Guastello, E. F. (2002). Concept mapping effects on science content comprehension of low-achieving inner-city seventh graders. *Remedial and Special Education, 21*(6), 356–364.

Harris, P. L. (1994). Thinking by children and scientists: False analogies and neglected similarities. In L. A. Hirschfeld & S. A. Gelman (Eds.), *Mapping the mind* (pp. 294–315). Cambridge, MA: Cambridge University Press.

Hewson, P. W. (1981). A conceptual change approach to learning science. *European Journal of Science Education, 3,* 383–396.

Hewson, P. W. (1982). A case study of conceptual change in special relativity: The influence of prior knowledge in learning. *European Journal of Science Education, 4,* 61–78.

Horton, S. V., Lovitt, T. C., & Bergerud, D. (1990). The effectiveness of graphic organizers for three classifications of secondary students in content area classes. *Journal of Learning Disabilities, 23*(1), 12–22.

Hughes, D., & Schumaker, J. B. (1991). Test-taking strategy instruction for adolescents with learning disabilities. *Exceptionality, 2,* 205–221.

Keraro, F. N., Wachanga, S. W., & Orora, W. (2007). Effects of cooperative concept: Mapping teaching approach on secondary school students motivation in biology in Gucha District, Kenya. *International Journal of Science and Mathematics Education, 5,* 111–124.

Lawrence-Brown, D. (2004). Differentiated instruction: Inclusive strategies for standards-based learning that benefit the whole class. *American Secondary Education, 32*(3), 34–62.

Libarkin, J., & Asghar, A. (2002). *How children think about light and invisible light.* Paper presented at the National Association for Research in Science Teaching (NARST) Conference, New Orleans, Louisiana. April 7, 2002.

Lindstrom, C., & Sharma, M. D. (2009). Link maps and map meetings: Scaffolding student learning. *Physics Review Special Topic–Physics Education Research, 5*(1), 1–11.

Lovitt, T. C., & Horton, S. V. (1994). Strategies for adapting science textbooks for youth with learning disabilities. *Remedial and Special Education, 15*(2), 105–116.

Markow, P. G., & Lonning, R. A. (1998). Usefulness of concept maps in college chemistry laboratories: Students' perceptions and effects on achievement. *Journal of Research in Science Teaching, 35*(9), 1015–1029.

Mastropieri, M. A. (2006). Differentiated curriculum enhancement in inclusive middle school science. *The Journal of Special Education, 40*(3), 130–137.

Mastropieri, M. A., Scruggs, T. E., & Graetz, J. (2005). Cognition and learning in inclusive high school chemistry classes. In T. E. Scruggs & M. A. Mastropieri (Eds.), *Advances in learning and behavioral disabilities: Cognition and learning in diverse settings* (pp. 107–118). Oxford, UK: Elsevier.

Mastropieri, M. A., Scruggs, T. E., & Levin, J. R. (1987). Learning disabled students' memory for expository prose: Mnemonic vs. non-mnemonic pictures. *American Educational Research Journal, 24,* 505–519.

Mastropieri, M. A., Scruggs, T. E., Norland, J. J., Berkeley, S., McDuffie, K., Tornquist, E. H., & Connors, N. (2006). Differentiated curriculum enhancement in inclusive middle school science: Effects on classroom and high-stakes tests. *Journal of Special Education, 40*(3), 130–137.

Mastropieri, M. A., Scruggs, T. E., & Shiah, S. R. L. (1991). Mathematics instruction with learning disabled students: A review of research. *Learning Disabilities Research and Practice, 6,* 89–98.

Maxwell, J. (2005). *Qualitative research design: An interactive approach.* London: Sage Publications.

McCarthy, C. B. (2005). Effects of thematic-based, hands-on science teaching versus a textbook approach for students with disabilities. *Journal of Research in Science Teaching, 42,* 245–263.

McCloskey, M., & Kargon, R. (1988). The meaning and use of historical models in the study of intuitive physics. In S. Strauss (Ed.), *Ontogeny, phylogeny and historical development* (pp. 49–67). Norwood, NJ: Ablex.

McLeskey, J., Rosenberg, M. S., & Westling, D. (2010). *Inclusion: Effective practices for all students.* New York: Pearson.

Miles, M., & Huberman, M. (1994). *Qualitative data analysis. An expanded sourcebook.* London: Sage Publications.

Moore, D. W., & Readence, J. F. (1984). A quantitative and qualitative review of graphic organizer research. *Journal of Educational Research, 78,* 11–17.

Nakhleh, M. B., & Samarapungavan, A. (1999). Elementary school children's beliefs about matter. *Journal of Research in Science Teaching, 36*(7), 777–805.

Nelson, J. R., Smith, D. J., & Dodd, J. M. (1992). The effects of teaching a summary skills strategy to students identified as learning disabled on their comprehension of science text. *Education and Treatment of Children, 15*(3), 228–243.

Novak, J. D. (1990). Concept maps and vee diagrams: Two metacognitive tools for science and mathematics education. *Instructional Science, 19,* 29–52.

Novak, J. D. (1991). Clarify with concept maps: A tool for students and teachers alike. *The Science Teacher, 58,* 45–49.

Novak, J. D. (1993). Human constructivism: A unification of psychological and epistemological phenomena in meaning making. *International Journal of Personal Construct Psychology, 6,* 167–193.

Novak, J. D. (1998). *Learning, creating, and using knowledge: Concept maps as facilitative tools in schools and corporations.* Mahwah, NJ: Lawrence Erlbaum Associates.

Nussbaum, J. (1985). The particulate nature of matter in the gaseous state. In R. Driver, E. Guesne, & A. Tiberghien (Eds.), *Children's ideas in science* (pp. 145–169). Milton Keynes, UK: Open University Press.

Osborne, R. (1980). *Force: Learning in science project.* Hamilton, New Zealand: University of Waikato.

Perkins, D. (1992). *Smart schools: From training memories to educating minds.* New York: Free Press.

Piaget, J. (1973). *The child's conception of the world.* London: Paladin.

Piaget, J., & Inhelder, B. (1974). *The child's construction of quantities: Conservation and atomism* (A. J. Pomerans, Trans.). London: Routledge and Kegan Paul.

Roberts, V., & Joiner, R. (2007). Investigating the efficacy of concept mapping with pupils with autistic spectrum disorder. *British Journal of Special Education, 34*(3), 127–135.

Rosenberg, M., Westling, D., & McLeskey, J. (2008). *Special education for today's teachers: An introduction.* Upper Saddle River, NJ: Pearson.

Sabornie, E. J., & deBettencourt, L. U. (2009). *Teaching students with mild and high-incidence disabilities at the secondary level* (3rd ed.). Upper Saddle River, NJ: Pearson Merrill Prentice Hall.

Samarapungavan, A., & Wiers, R. (1997). Children's thoughts on the origin of species: A study of explanatory coherence. *Cognitive Science, 21,* 147–177.

Schneps, M. H., & Sadler, P. (1997). *Minds of our own.* Boston, MA: Harvard Center for Astrophysics.

Scruggs, T. E., Mastropieri, M. A., Bakken, J. P., & Brigham, F. J. (1993). Reading vs. doing: The relative effectiveness of textbook-based and inquiry-oriented approaches to science education. *The Journal of Special Education, 27,* 1–15.

Scruggs, T. E., Mastropieri, M. A., Berkeley, S., & Graetz, J. E. (2010, November/December). Do special education interventions improve learning of secondary content? A meta-analysis. *Remedial and Special Education, 31,* 437–449.

Scruggs, T. E., Mastropieri, M. A., & Okolo, C. M. (2008). Science and social studies for students with disabilities. *Focus on Exceptional Children, 41*(2), 1–24.

Seaman, T. (1990). *On the high road to achievement: Cooperative concept mapping.* (ERIC Document Reproduction Service No. ED 335 140)

Sindelar, P. T., Monda, I., E., & O'Shea, I. J. (1990). Effects of repeated readings on instructional- and mastery-level readers. *Journal of Educational Research, 83,* 220–226.

Skylar, A. A., Higging, K., & Boone, R. (2007). Strategies for adapting WebQuests for students with learning disabilities. *Intervention in School & Clinic, 43*(1), 20–28.

Smith, C. (1988). *Weight, density and matter: A study of elementary children's reasoning about density with concrete materials and computer analogs.* Cambridge, MA: Harvard Graduate School of Education, Educational Technology Center.

Smith, C., Maclin, D., Grosslight, L., & Davis, H. (1997). Teaching for understanding: A study of students' preinstruction theories of matter and a comparison of the effectiveness of two approaches to teaching about matter and density. *Cognition and Instruction, 15*(3), 317–393.

Smith, S. J., & Okolo, C. (2010). Response to intervention and evidence-based practices: Where does technology fit? *Learning Disability Quarterly, 33*(4), 257–272.

Snow, C. (2010). Academic language and the challenge of reading for learning about science. *Science, 5977*(328), 393–532.

Stavy, R., & Stachel, D. (1984). *Children's ideas about solid and liquid.* Tel Aviv: Tel Aviv University, Israel Science Teaching Center.

Stepans, J. (2003). *Targeting students science misconceptions.* Tampa, FL: Showboard.

Strauss, A., & Corbin, J. (1998). *Basics of qualitative research: Grounded theory, procedures and techniques.* Newbury Park, CA: Sage.

Strike, K. A., & Posner, G. J. (1985). A conceptual change view of learning and understanding. In L. H. West & A. L. Pines (Eds.), *Cognitive structure and conceptual change* (pp. 211–232). London: Academic Press.

Strike, K. A., & Posner, G. J. (1992). A revisionist theory of conceptual change. In R. Duschl & R. Hamilton (Eds.), *Philosophy of science, cognitive psychology, and educational theory and practice* (pp. 147–176). Albany, NY: SUNY Press.

Thagard, P . (1989). Explanatory coherence. *Behavioral and Brain Science, 12,* 435–502.

Vosniadou, S. (1994). Capturing and modeling the process of conceptual change. *Learning and Instruction, 4,* 45–69.

Vosniadou, S. (2007). The cognitive-situative divide and the problem of conceptual change. *Educational Psychologist, 42*(1), 55–66.

Vosniadou, S., & Brewer, W. F. (1992). Mental models of the Earth: A study of conceptual change in childhood. *Cognitive Psychology, 24,* 535–585.

Waldron, N., & McLeskey, J. (2001). An interview with Nancy Waldron and James McLeskey. *Intervention in School and Clinic, 36*(3), 175–182.

Webb, P. (2010). Science education and literacy: Imperatives for the developed and developing world. *Science, 5977*(328), 448–450.

Wiser, M. (1988). The differentiation of heat and temperature: History of science and novice-expert shift. In S. Strauss (Ed.), *Ontogeny, phylogeny, and historical development* (pp. 28–48). Norwood, NJ: Ablex.

Woodward, J. (1994). The role of models in secondary science instruction. *Remedial and Special Education, 15*(2), 94–104.

Yates, G. C. R., & Yates, S. (1990). Teacher effectiveness research: Towards describing user-friendly classroom instruction. *Educational Psychology, 10,* 225–238.

CHAPTER 3

THE METACOGNITIVE SCIENCE TEACHER

A Statement for Enhanced Teacher Cognition and Pedagogy

Gregory P. Thomas

INTRODUCTION

Setting the Scene on an Understudied Area of Science Education

In the invitation to contribute to this book, prospective authors were asked to consider several foci regarding contemporary science teaching approaches. These included that students should be fostered to construct scientific knowledge as a process of inquiry rather than as passive learners, and that science teaching approaches should enhance students' conceptual understanding of scientific concepts. Also outlined was how the book would develop a deeper knowledge of current best practice in science education and introduce contemporary pedagogical approaches in science teaching. The position taken in this chapter is that a substantial under-

Contemporary Science Teaching Approaches, pages 29–53
Copyright © 2012 by Information Age Publishing
29

standing of metacognition is an essential precursor for enabling teachers to engage in practices that are consistent with achieving student learning outcomes that are consistent with the aforementioned foci.

The arguments presented in the chapter rest on the following reasoning. Metacognition, and its development and enhancement, is a key to improving students' science learning. All students are metacognitive to some extent, and their metacognition develops as a consequence, at least in part, of their experiences within their classroom learning environments. Some students' metacognition is not adaptive for the demands of their learning environments. To improve students' science learning, there is a need to develop and enhance their adaptive metacognition so that they can learn science more effectively, efficiently, and with increased understanding across science learning contexts. To develop their metacognition, there is a need to make science learning environments more and differently metacognitively oriented. To do this, teachers need to understand metacognition, to be metacognitive themselves, and to understand and be able to implement pedagogies that facilitate students' metacognitive development. This means it is necessary to explicitly teach students the cognitive strategies required to learn science well and also how to self-manage the use of those strategies. This is in contrast to assuming that students will "automatically" do so, or that deficit models of cognition or ill-informed stage-based views can adequately explain students lack of success as science learners. Teacher metacognition is therefore an essential element of improving science teaching and learning.

A review of the literature suggests that very little has been written specifically about teacher metacognition in science education or in any other subject area. This might be for a number of reasons. Firstly, the focus of much metacognition research in science education has been on pedagogical interventions that have aimed to enhance students' cognition, metacognition, and learning outcomes. Secondly, an assumption seems to have been that teachers are inherently knowledgeable in relation to their own metacognition and cognition and also that of their students. This is not necessarily the case in many if not most instances, and this assumption needs to be critiqued. Thirdly, it may be that science teacher metacognition has been subsumed in other research such as that on teacher reflection, pedagogical content knowledge, and self-study.

The purpose of this chapter is to make science teacher metacognition a more explicit area for consideration by teachers, teacher educators, and researchers. The chapter is structured as follows: An overview of the nature of metacognition and its importance of science learning introduces readers unfamiliar with metacognition to this field of study. This is followed by a review of metacognitively oriented science classroom learning environments (MOSCLEs) and teachers' roles within those environments. These roles are in addition to but commensurate with those that have been more tradition-ally ascribed. Understanding teacher's roles within MOSCLEs establishes the centrality and importance of teacher metacognition as a factor that fa-cilitates and is necessary for the development of such environments. It is ar-gued that teachers who are themselves not appropriately metacognitive in relation to their own science learning are less capable of effecting change in students in relation to their learning processes and metacognition, and that a first step to enabling teachers to do so is to provide opportunities for teachers to develop their own metacognition. An answer to the question, "What might science teachers' metacognition 'look' like?" is explored in conjunction with a review of available literature on teacher metacognition in general and science teacher metacognition in particular. The chapter concludes with suggestions for research in this potentially fertile area.

Metacognition and Its Importance for Science Learning

Research into metacognition within science education has a history of around 30 years. It can be traced to the excitement resulting from the pi-oneering work of Flavell (1971, 1976, 1979) and Brown (1978), who are largely credited with the initial conceptualization/s of the concept. Rather than attend to issues of definition and methodology (e.g., Dinsmore, Al-exander, & Loughlin, 2008; Lajoie, 2008), those conducting research on metacognition in science education have typically looked to apply meta-cognitive theory to classroom learning environments, including laborato-ries (e.g., Davidowitz & Rollnick, 2003) and, more recently, informal learn-ing environments (e.g., Anderson, Thomas, & Nashon, 2009; Thomas, Anderson, Ellenbogen, & Cohn, 2009). They have sought to improve stu-dents' learning science by developing and enhancing their metacognition (e.g., Baird, 1986; Blank, 2000; Connor, 2007; Georghiades, 2006; Thomas & McRobbie, 2001). This trend can be explained at least in part by under-standing that many researchers involved in metacognition in science edu-cation have themselves been, and in some cases continue to be, classroom science teachers. Consequently, they are interested in improving classroom practice and student learning. The author of this chapter is a product of that tradition. It is clear from scholarship in science education that meta-cognition is a key element of science learning and that enhancing students' metacognition can enhance their science learning (e.g., Donovan & Brans-ford, 2005; Georghiades, 2004; Thomas, in press; White, 1998; White & Frederiksen, 1998; White & Mitchell, 1994).

The precise definition of metacognition is still a subject of debate (Garner & Alexander, 1989; Hacker, 1998; Veenman, Van Hout Wolters, & Afflerbach, 2006), and it is not the intention of this author to engage in this debate. Metacognition is defined in this chapter as an individual's knowledge, and conscious (Borokowski, Carr, & Pressley, 1987) control and awareness, which implies monitoring, of their thinking and learning processes. Also, an individual's knowledge of how others think and learn might be considered elements of their metacognition, because such knowledge can act as a referent that informs their own thinking and learning processes. It is important to realize that a clear distinction needs to be made between metacognition and cognition even though the two are closely related (Nelson & Narens, 1994). There is often confusion in distinguishing between cognition and metacognition, and some examples relevant to science education can illustrate this point and clarify this distinction.

A cognitive process in science learning closely related to constructivist conceptual change perspectives is accessing one's existing knowledge of a concept and comparing that existing knowledge with new information that is either presented by a teacher or that arises via some other learning experience, for example a laboratory experiment or a field trip. There is no doubt that such a cognitive process is valuable for effective science learning. The metacognitive higher-order relative of this cognitive process is the conscious awareness that one has knowledge of this process, control over the use of this process, and the awareness of when one is doing so and why doing so is beneficial. Another cognitive process that is employed commonly by science learners is that of memorization. There is no doubt that there are times when memorization of science material is highly valuable for science learning and the development of science understanding, even if its importance is lost these days on some science teachers and science teacher educators. An effective science learner will possess cognitive strategies for memorizing science material that they consider to be important or that they are told is important. These strategies may include the use of acronyms and mnenonics. As with the previous example, an effective and metacognitive science learner will have knowledge of (a) memorization as a learning strategy, (b) the affordances and constraints of memorization, and (c) how and when it might be best undertaken. They would also have conscious control of their memorization strategies and awareness of when they employ such strategies. What is common between these two examples is that there is clearly a metacognitive knowledge base for metacognition that relates to science learning as shown in Figure 3.1. Figure 3.1 helps set a foundation for discussing metacognitive knowledge.

Figure 3.1 suggests that individuals enact their metacognition to engage in cognitive and learning processes to produce an outcome. Ideally, reflection on and evaluation of the cognitive and learning processes occurs so the metacognition can be modified to improve it. The metacognition (meta-level) is superordinate to the cognitive and learning processes (object-level).

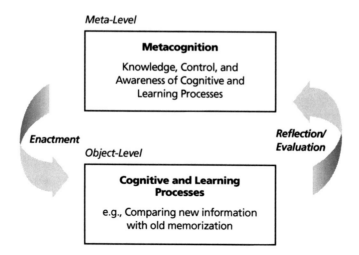

Figure 3.1   Metacognition and cognition: A relationship (student focus). (Adapted from Nelson & Narens, 1994)

The Centrality of Metacognitive Knowledge

Metacognitive knowledge consists of declarative, procedural, and conditional knowledge elements. The declarative elements include beliefs, opinions, theories, hypotheses, conceptions, and facts about what learning (science) is. Procedural metacognitive knowledge includes an individual's knowledge of information storage, retrieval and processing strategies, and he or she knowing how to perform cognitive and learning activities, such as the two aforementioned cognitive strategies of comparing new information with already existing knowledge and memorization. Declarative and procedural knowledge interact. For example, students' conceptions of learning (declarative knowledge) often determine how they accomplish learning (procedural knowledge). Having declarative knowledge of learning strategies is typically a prerequisite for successfully implementing procedures of learning. Students might have difficulty in learning in a particular subject area, for example, physics, because they may lack specific knowledge of the subject material or because they do not understand the requisite procedures required to learn and understand that material. Discovering which is more deficient is a necessary step in planning effective and/or remedial instruction.

Conditional knowledge relates to knowing when to employ various types of declarative and procedural metacognitive knowledge and why it is important to do so. Just knowing facts and procedures regarding what science learning is and how it can be achieved is insufficient to achieve high-quality science learning (Boekaerts, 1997). It is students' conditional knowledge, including understanding of both the value and limitations of learning strategies and knowing when, how, and why to apply those strategies that dictates whether or not they will be purposefully and appropriately employed. An individual's metacognitive knowledge regarding their learning performance in science subjects would, therefore, include knowledge of his/her strengths, weaknesses, and learning processes, together with an awareness of his/her repertoire of tactics and strategies and how and under what circumstances these could enhance or inhibit learning and/or cognitive performance. Figure 3.2 illustrates a taxonomy for metacognitive knowledge.

This taxonomy supports the notion that the development of students' metacognition may be facilitated by developing their metacognitive knowledge of learning and learning processes that are appropriate for their learning of science. This is not to say that students do not and cannot develop such metacognitive knowledge related to science learning without explicit teacher assistance or formal guidance. Their very presence in science classrooms means that teachers and other students either explicitly, or tacitly more likely, convey to them what it means to learn and understand science, what is valued as science learning, and how learning science should be undertaken. This view is consistent with a social cognition perspective on learning that holds that every person develops understanding and constructs knowledge within the context of a culture. Their culture and the type of thinking and learning that is valued within it shape what learning they perceive as important and what processes they undertake to achieve their culturally mediated learning goals (Donovan & Bransford, 2005; Thomas, 2002).

Therefore, all students will have metacognitive knowledge in one form or another as suggested by Gunstone (1994). Research repeatedly confirms this point. A primary concern with this realization is that students' metacognition and learning processes might not always be adaptive for the demands of their contexts, especially as they progress to science classes with more-advanced content and process demands, and with higher con-

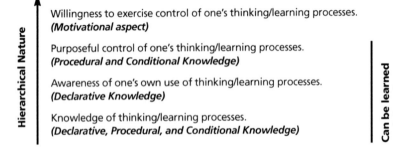

Figure 3.2 A taxonomy for metacognitive knowledge.

ceptual learning requirements. For example, memorization may be perceived by students to be a viable and principal strategy in some science learning contexts, especially in those in which assessment consists predominantly of low-cognitive-demand recall questions. If they are not exposed to experiences in which higher-order thinking, problem solving, and/or authentic inquiry are called for, it is possible that they will conceive of science learning as a predominantly low-cognitive-demand endeavor relying largely on memorization. This possibility will be increased if they are successful in meeting assessment demands and achieving high grades in such contexts that, in essence, trivialize science learning. Thomas (1999) and Thomas and McRobbie (2001) reported on Debbie, a Year 11 chemistry student who had been successful in junior high but who increasingly struggled to learn science as the material became more complex than that of junior high and as assessment demands changed. Debbie had come to believe that "working hard" and "efficiency" were important for learning success, and that repetition of algorithmic processes with a view to meeting what she wrongly considered the demands of the chemistry assessment was also a key process. With a change in the increasing cognitive and assessment demands in science over the latter years of secondary school, Debbie and other members of her class experienced metacognitive conflict (Thomas, in press) when they realized that (a) their existing learning processes were not adequate for the level of learning that was required and (b) they needed to consider alternative strategies and beliefs about learning science if they were to increase their achievement. Employing a taxonomy based on declarative, procedural, and conditional elements of metacognitive knowledge, it is possible to begin to evaluate the quality and quantity of students' metacognition (as was done with Debbie) according to the nature of their metacognitive knowledge. From there it is necessary to ascertain whether students' metacognitive knowledge is adaptive and if and how they employ their knowledge to achieve learning goals through its purposeful, conscious application and control.

There are many students like Debbie in science classes worldwide. They possess metacognitive knowledge, but it is often not adaptive for higher order thinking or for the demands of learning more complex material in the middle to latter years of secondary school and beyond. Further, their metacognitive knowledge is often tacit and difficult for them to verbalize or describe (Thomas, 2010a; Thomas & McRobbie, 1999). It is essential that students be explicitly taught the thinking strategies and processes required for effective and successful science learning. For this to occur, they need to learn and to share a common language, a language of thinking and learning (e.g., Macdonald, 1990; Thomas & McRobbie, 2001; Tishman & Perkins, 1997; Zohar, 2004) with teachers and other students regarding those strategies and processes. Therefore, unless teachers are knowledgeable about the thinking processes required to learn science, can and do make expectations and required cognitive processes explicit for students, and model those cognitive processes, then the chances of students learning and managing those processes are diminished.

If we accept that there is a body of metacognitive knowledge that is valuable to possess for enabling effective science learning, then we might also accept that such metacognitive knowledge can be explicitly taught to students as suggested above. Schraw (1998) previously proposed this possibility, although it did not receive great support at the time (e.g., Sternberg, 1998). On closer review, however, and in light of new research findings (e.g., Thomas & Au, 2005; Thomas & McRobbie, 2001; Yürük, 2005), it appears that Schraw's proposal may indeed be viable. If this is the case, then metacognitive knowledge needs to be developed in students in an explicit and overt manner, something that is not currently done or evident in the vast majority of science classrooms. This has implications for what teachers need to know regarding how students learn science and in terms of their own metacognitive knowledge. Teachers are major determinants of what occurs in classrooms. Consequently, it is only reasonable to expect that if we want students to develop and enhance metacognition and cognitive strategies that are adaptive for higher-order thinking and for developing a comprehensive understanding of science and its processes, then we would also expect that teachers would take a lead role in constructing learning environments that are metacognitively oriented. It is therefore appropriate at this point to explore what is meant by a metacognitively-oriented science classroom learning environment.

Metacognitively-Oriented Science Classroom Learning Environments (MOSCLEs): An Overview

Thomas (2003, 2004) conceptualized the existence and confirmed the factorial validity of seven dimensions that can be used to ascertain the extent to which science classroom learning environments are metacognitively oriented. Table 3.1 summarizes these dimensions.

It is clear from Table 3.1 that the teacher is a central figure in developing MOSCLEs. In MOSCLEs, teachers, for example, (a) ask students to be aware of how they learn and how they can improve their science learning, (b) discuss students' science learning processes with them, and (c) encourage students to improve their science learning processes. These classroom environment dimensions are reflected in the Metacognitive Science Learning Environment Scale–Science (MOLES-S) (Thomas, 2003), a quantitative learning environments instrument used to determine the extent to which science classrooms are metacognitively oriented (e.g., Thomas, 2006,

TABLE 3.1   Dimensions for Metacognitively-Oriented Science Classroom Learning Environments

Dimension Name	Dimension Description
Metacognitive Demands	Students are asked by the teacher to be aware of how they learn and how they can improve their science learning.
Student-Student Discourse	Students discuss their science learning processes with each other.
Student-Teacher Discourse	Students discuss their science learning processes with their teacher.
Student Voice	Students feel it is legitimate to question the teacher's pedagogical plans and methods.
Distributed Control	Students collaborate with the teacher to plan their learning as they develop as autonomous learners.
Teacher Encouragement & Support	Students are encouraged by the teacher to improve their science learning processes.
Emotional Support	Students are cared for emotionally in relation to their science learning.

2010b). It is a valid and reliable tool that provides some, but incomplete, insights into the metacognitive orientation of science classroom learning environments. Ideally, data collected and assertions regarding arising from the use of the MOLES-S should be triangulated against data and assertions arising through the use of other methods such as interviews and long-term classroom observations.

One other characteristic that was considered initially by Thomas (2003) to be essential in MOSCLEs was the aforementioned teacher modeling of cognitive strategies and science learning processes. Items reflecting this characteristic did not survive the stringent factor analysis used to develop the MOLES-S. This does not mean teacher modeling should not be considered. Effective teaching of metacognitive (Donovan & Bransford, 2005; Schraw, 1998) and cognitive (Lajoie, 2008) strategies relies on initial teacher modeling with a gradual shift to students. The suggestion that teacher modeling is important is commensurate with the view that students learn from observing and copying the behaviors, cognitive strategies, and thinking processes of successful learners, not just in science learning but across all spheres of experience. Indeed, at the high performance levels of sport and business, considerable attention is given to enhancing individuals' cognitive and metacognitive domains with a view to maintaining high performance. Unfortunately, this seems not to be the case in many science classrooms where teachers are often not appropriate metacognitive role models for their students, as will be discussed later. It could be argued that this is in large part because of teachers' lack of metacognition in relation to their own science learning processes, their lack of understanding of the importance of their metacognitive knowledge, and their lack of understanding of how to act as cognitive and metacognitive role models for their students and the importance of doing so.

I stress that this role entails being more than an exciting and passionate science teacher. It means more than asking students just to use concept maps, venn/relational diagrams (White & Gunstone, 1992), and/or Vee diagrams (e.g., Mintzes & Novak, 2005). It means more than developing students' abilities to hypothesize, to analyze data and come to conclusions on the basis of evidence. It means more than developing students' capacity to balance redox equations, or to solve chemical equilibrium, Newtonian physics, or dihybrid cross-problems. It means more than getting students to do activities like Predict-Observe-Explain (POE) (White & Gunstone, 1992), which can lead to science learning. It means that teachers need to elaborate and model a view of what it means to learn and understand science and how to learn science with understanding. This is not to discount the importance of the aforementioned activities and related cognition, but students need to be "educated" in terms of more than just the science material. If they are to be asked to use "metacognitive activities" (Georghiades, 2006; Thomas, in press), such as those used by Georghiades (2006), Connor (2007), Davidowitz and Rollnick (2003), and Blank (2000), then they should also be asked and supported to construct metacognitive knowledge and understanding regarding the nature of these strategies by fully considering the declarative, procedural, and conditional claims and principles associated with the use of those strategies. This is a realistic expectation.

So, and importantly, it is necessary to consider the qualitative nature of a classroom environment's metacognitive orientation in addition to its quantitative character. Even if teachers are asking students to use metacognitive activities to learn science, there is no guarantee that they are asking them to be aware of how these activities help them to learn science. Even if teachers are asking students to be aware of how they learn and how they can improve their science learning, there is no guarantee that they are asking them to consider using alternative, possibly more adaptive learning strategies and cognitive processes. Even if teachers are discussing students' science learning processes with them, there is no guarantee that they are discussing the learning processes and cognitive strategies that will result in the type of science learning that reflects contemporary science education goals. Science teachers most often focus on the science material in science lessons and not on how the science can be best learned and/or understood. Further, science classrooms are not particularly metacognitively oriented either quantitatively or qualitatively. This is hardly surprising given that science teachers' primary role is to teach THE science (author's emphasis).

Currently, enacted science curricula tend to predominantly inform the science content and the understanding and related, predominantly algorithmic processes to be learned by students to meet systemic assessment requirements. What is needed is a metacognitively-oriented curriculum that complements and runs concurrently within existing science curricula, as suggested by Perkins (1992) and Zohar (2004). Unfortunately, these suggestions have been almost completely ignored in most education circles. The metacognitively-oriented curriculum would focus on the development of students' adaptive declarative, procedural, and conditional metacognitive knowledge. Just as science teachers are expected to weave nature of science and science, technology and society themes into their day-to-day science teaching, they should also be expected to explicitly weave the development of students' adaptive metacognition into their science teaching. This would require a paradigm shift in science education circles that continue to be focused on, for example, conceptual change, nature of science, STSE, and scientific inquiry. However, Thomas (2009a) explained how a metacognitive agenda in science education is absolutely compatible with these long-standing agendas. This position is supported by the work of Abd-El-Khalik and Akerson (2009) in relation to the nature of science. They argued that "training in metacognitive strategies substantially enhanced the effectiveness of explicit-reflective NOS instruction in impacting prospective elementary teachers' conceptions of NOS" (p. 2180). It is time to reconsider how metacognition can be attended to more explicitly in science curricula and classrooms.

Therefore, it is prudent to explore teacher characteristics that would be necessary to enable such a paradigm shift to become a reality. Teacher metacognition would be high on any list of enabling teacher characteristics. Teachers of science are expected to have substantial knowledge and deep understanding of the subject material they teach and to be or have been successful science learners. They are also expected to have well-developed and high-quality pedagogical content knowledge, to encourage student participation, and to have considerable skills of classroom management. To be able to develop students' metacognition, they should also have a substantial understanding of the cognitive processes and strategies required to learn science well, and they should be able to articulate such understandings. These understandings should be consistent with current learning theory as it relates to science learning and the various agendas in science education. They should also be metacognitive themselves (Zohar, 2004) and able to articulate their own metacognitive knowledge and thought processes to students. The point being made here is this: if teachers are to teach students to be adaptively metacognitive so they can meet higher-order demands of their science learning environments, then teachers themselves need to be metacognitive and able to discuss and model learning and thinking processes with and for students. In the next section, teacher metacognition is explored in more depth. The section begins with exploring further the question, "What does teacher metacognition 'look' like?" Incorporated is a review of the existing, but quite meager literature that currently exists on science teacher metacognition and teacher metacognition in general. Implications are drawn from this discussion.

"WHAT DOES SCIENCE TEACHER METACOGNITION 'LOOK' LIKE, AND WHAT DO WE KNOW ABOUT IT?"

If we consider the general definition of metacognition previously referred to, then a definition of teacher metacognition might logically be, "A teacher's knowledge, control, and awareness of his/her own thinking and learning processes as they relate to science teaching and learning." This includes and is related to science teachers' thinking and learning about a diverse range of matters associated with science teaching and learning. So what might be included according to this definition? What do we want teachers to be metacognitive about? What cognition do we want them to have knowledge, control, and awareness of? To answer these questions means that we should consider not only the type of metacognition that is related to the development of conceptual understanding of science material or problem solving, even though this is of great importance and an ongoing focus as suggested in this chapter's first paragraph. Teachers' work is multifaceted, and other thought and learning processes and strategies are also important. Therefore, also included in their metacognition would be metacognitive knowledge associated with making curricular and pedagogical decisions.

Table 3.2 provides an overview of three categories of metacognitive knowledge that might form a foundation of science teachers' metacognition that one might expect them to have conscious, purposeful control and awareness of. These categories are by no means exhaustive regarding all metacognitive knowledge that we might expect science teachers to possess, to be able to articulate, and to model. They do however reflect knowledge related to a selection of cognitive "object level" processes that are generally acknowledged as being important for science teaching. Figure 3.3 is an adaptation of Figure 3.1 illustrating the relationship between the three aforementioned teacher thinking processes and the meta-level knowledge control and awareness of those processes.

Scientific/Conceptual-Oriented Metacognitive Knowledge (S/CO-MK) is given more attention in what follows because it relates specifically to the aforementioned foci of this book regarding enhancing students' conceptual understanding of science concepts. Curricular-Oriented Metacognitive Knowledge (CO-MK) and Pedagogically-Oriented Metacognitive Knowledge (PO-MK) are attended to collectively as it could be argued that they have been the source of deliberation and research in science education

TABLE 3.2 Three Categories of Essential Teacher Metacognitive Knowledge

Category Name	Category Description	Example
Scientific/Conceptual-Oriented Metacognitive Knowledge (S/CO-MK)	Teachers' knowledge of their own thinking and science learning processes and strategies.	Knowledge of the use of memorization and visualization strategies for advancing their own learning.
Pedagogically-Oriented Metacognitive Knowledge (PO-MK)	Teachers' knowledge of how they think and what strategies they employ when making decisions regarding which pedagogical approaches to employ.	Knowledge of how to make reasoned selection between several feasible pedagogies for teaching a topic.
Curricular-Oriented Metacognitive Knowledge (CO-MK)	Teachers' knowledge of how they think and what strategies they employ when making curricular decisions regarding which emphases to focus on in their teaching.	Knowledge of how to make reasoned selection between two curriculum emphases.

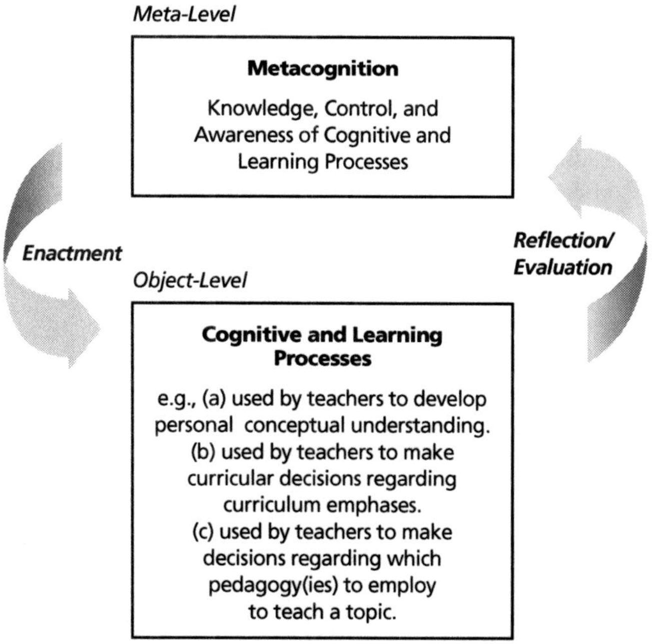

Figure 3.3 Metacognition and cognition: A relationship (student focus). (Adapted from Nelson & Narens, 1994)

for some time, even if a metacognition framework has not been applied to such scholarship. It is suggested that employing a metacognition framework is a useful way of conceptualizing teachers' thinking in relation to their pedagogical and curricular cogitation/s.

SCIENTIFIC/CONCEPTUAL-ORIENTED METACOGNITIVE KNOWLEDGE (S/CO-MK)

A science teacher's own knowledge about how he/she learns and learned science—the strategies that were successful and those that were not, and when and why they are or are not and were or were not successful—should obviously be included as essential metacognitive knowledge. As Schraw (1998) suggests, "The starting point for this endeavour is for teachers (or expert students) to ask themselves what skills and strategies are important within the specific domain they teach, how they constructed those skills, and what they can tell their students about using those skills intelligently" (p. 123). Ideally, it would also relate strongly to teachers' self-assessment of their own learning practices with reference to the existing literature regarding (a) the processes individuals use to learn science, (b) the difficulties that individuals experience learning science, (c) the alternative conceptions they construct, and (d) the poor student-learning tendencies that are continually evident. Such knowledge is crucial if teachers are to inform students about how they might best learn science, and so teachers can act as cognitive and metacognitive role models for students. S/CO-MK would encompass teachers' procedural and conditional metacognitive knowledge of their learning strategies and processes. A teacher's own beliefs about what it means to learn, to know, and to understand science, and what science knowledge and understanding "is" would also be included in their S/CO-MK. These beliefs might be categorized, at least in part, as declarative epistemic metacognitive knowledge (Mason, Boldrin, & Ariasi, 2010). It could be expected that much of the declarative epistemic metacognitive knowledge of teachers would vary according to their academic and cultural backgrounds and experiences.

Studies focusing on this S/CO-MK element of science teacher metacognition have already, albeit relatively recently, been undertaken (e.g., Leou, Abder, Riordan, & Zoller, 2006; Zohar, 1999, 2004). These studies are located within the contexts of science teacher professional development programs in Israel and the United States, aimed to increase the "capacity" of teachers to implement educational reforms. Zohar's writings are particularly noteworthy because of the magnitude of the research they are drawn from and because of their contributions to emerging understandings in this area. The professional development in both cases targeted certain science-learning-specific thinking strategies and then explored the teachers' metacognition in relation to these strategies. Embedding metacognitive training within subject-specific contexts was seen as most appropriate (Donovan & Bransford, 2005; Gunstone, 1994). In the Zohar research, these strategies related to the infusion of higher order thinking into the science curriculum rather than as a separate subject. In the Leou and colleagues study, those strategies related to the higher-order thinking strategies (HOTS) of question asking, problem solving, and conceptualization of fundamental concepts in the context of helping teachers making more effective use of nonformal science learning environments.

These studies found that (a) the teachers' intuitive (i.e., pre-instructional) knowledge of metacognition and thinking skills is/was unsatisfactory for the purpose of teaching higher-order thinking skills in science classrooms; (b) teachers have trouble explicitly describing and explaining their thinking and learning processes, that is, they often lack a language of thinking and therefore have difficulty expressing their metacognitive knowledge of thinking and learning strategies/skills; and (c) making such knowledge evident and explicit was a key element in the teachers' professional development, as evidenced by the following teachers' quotes from Zohar (2004):

> I think I went through a real process. I think that my awareness of things was sharpened a great deal. I think that a large part of what we got here—a large part of it we were doing intuitively—but making it conscious—I think this is the greatest thing I gained. (p. 189)

> Even the skills we usually practice in class, we are not conscious about it in the same way. We don't call them by names. (p. 189)

The finding that science teachers had difficulty expressing their metacognitive knowledge of thinking skills is consistent with that of Kozulin (2005), who reported from a non–science education, Instrumental Enrichment (IE) (Feuerstein, Rand, Hoffman, & Miller, 1980) context that strategy descriptions provided by many educators even after some level of IE training were "vague, inconsistent or irrelevant" (p. 3). Kozulin makes the following observations that, on the basis of the studies by Leou et al. (2006) and Zohar (1999, 2004), seem equally applicable to the situation in science education:

> If one accepts that teachers are expected to be particularly skilled in analyzing the problem solving (and learning) processes of their students, the fact of their difficulty in reflecting on their own problem solving indicates that the metacognitive aspect was severely neglected in their previous professional training. (p. 6)

> There is a considerable gap between teachers' ability to solve cognitive tasks and their ability (and disposition) to reflect upon their own problem solving strategies. (p. 6)

Within science education, these views are echoed by Zohar and Schwartzer (2005), who summarize the work of Zohar (1999, 2004), Zohar and Nemet (2002), and Zohar, Vaaknin, and Degani (2001), suggesting, "the findings of these studies show that most teachers do not have the metacognitive knowledge that is necessary for sound teaching of higher-order thinking" (p. 1599).

An implication of this is that science teachers need opportunities to make their S/CO-MK, their own thinking strategies, and metacognitive knowledge related to the conceptual science learning explicit to themselves, their peers, and to other education professionals. This doesn't often happen within preservice or in-service teacher education. A teacher's ability to apply thinking and learning strategies is a necessary precondition, but by itself, it is insufficient to "qualify" him/her to teach such processes and strategies. Overt consciousness of thinking and learning skills/strategies as potential learning goals and/or explicit instructional objectives is necessary for designing and implementing classroom activities to promote metacognition, higher-order thinking, and science learning. Explicit and reportable S/CO-MK (declarative, procedural, and conditional) is absolutely necessary.

On a positive note, the studies of Leou et al. (2006) and Zohar (1999, 2004) provide evidence that with encouragement and informed professional development, science teachers can begin to reflect on and explore their own learning processes and cognitive strategies and become more metacognitive and metacognitively knowledgeable. For example, teachers in Leou and colleague's study suggested,

> I began to think about the act of thinking and learning itself. As I reflect on my cognitive gains during the past 4 weeks, I realize that it is very important for us to ponder and think about thinking. We have to get into the minds of our students through higher levels of questioning. (p. 78)

> Forming HOCS (higher order cognitive skills) questions requires teachers also to become HOCS thinkers. I think this is great because teachers must develop these skills first before requiring them of their students. (p. 79)

Professional development for the purposes of developing teachers' metacognition should attend to making teachers' cognition and cognitive strategies objects of reflection and scrutiny. Teachers need to be able to articulate their own thinking as a precondition to increase their capacity to be able to develop sought-after thinking and learning strategies in their students.

PEDAGOGICALLY-ORIENTED METACOGNITIVE KNOWLEDGE (PO-MK) AND CURRICULAR-ORIENTED METACOGNITIVE KNOWLEDGE (CO-MK)

Science teachers' metacognitive knowledge regarding how they consider and make decisions regarding the pedagogies they employ—Pedagogically-Oriented Metacognitive Knowledge (PO-MK)—is a second category of essential teacher metacognitive knowledge. An example of the engagement of PO-MK might be as follows: A teacher has to teach an introductory lesson on chemical reactions at the lower secondary (junior high/middle school) level. A number of pedagogies are feasible and known to the teacher for teaching such a lesson. These possibilities might include a teacher-driven didactic lesson, a lab lesson, a demonstration lesson, an independent reading lesson with a series of questions, or a combination of any of these. The "object-level" cognition involved relates to the thinking process/es involved in deciding which pedagogy to employ. The metacognition involved relates to the teacher possessing and employing PO-MK regarding how such decisions can be and are made, being aware that one is employing such processes, and being in control of such processes. The actual lesson planning strategies are subordinate but related to the executive metacognitive processes.

How science teachers consider and make decisions regarding the science curriculum they implement—Curricular-Oriented Metacognitive Knowledge (CO-MK)—is a third category of essential teacher metacognitive knowledge. It is well known that teachers vary considerably in the curriculum emphases (Roberts, 1982, 1988) they choose in their teaching (Chu, 2009; Hepburn & Gaskell, 1998). These emphases include correct explanations, structure of science, and everyday coping. Teachers make decisions based on many factors such as available teaching time, their beliefs about what should be and what they consider is valued in science teaching and learning, and their perceptions of their own knowledge and expertise in relation to specific emphases. As an aside, following on from the previous reference to the metacognitively-oriented curriculum, we might add another emphasis that is superordinate to or infiltrates all other emphases: the emphasis of promoting metacognition and higher-order thinking in science education. As Zohar and Schwartzer (2005) suggest, "teachers must know how to implement a curriculum that addresses higher-order thinking goals, and/or how to plan lessons that are rich in such goals" (p. 1597). Following the framework in Figure 3.3, the cognitive processes a teacher employs to make decisions regarding which emphases to focus on, and to what extent, represent the cognitive or object level of engagement with this activity. The knowledge, conscious control, and awareness of the use of those processes represent the meta- or metacognitive level. Once again there is a need to stress the conscious nature of engagement at the metacognitive level. Teachers make such decisions on a daily basis, but the extent to which they are conscious of doing so and able to articulate their cognitive processes is a key consideration when evaluating their metacognition. When the processes become automated, they are no longer likely under metacognitive control. This may be useful to save time and reduce cognitive load, but it can be counterproductive if it leads to resistance to considering alternative, possibly more-productive thinking strategies regarding teaching and learning, or when contextual conditions change so that existing cognition is no longer adaptive for the changed context (Lin, Schwartz, & Hatano, 2005).

It would be absolutely wrong to suggest that there has been no research into science teacher thinking and how they make decisions. Much has been researched and written on such matters, and it is not possible in this chapter to review that literature (for exemplars see, Berry, Loughran, Smith, & Lindsay, 2009; Gess-Newsome & Lederman, 1999; Loughran, Berry, & Mulhall, 2006; van Driel, Verloop, & de Vos, 1998). Obviously, there is considerable overlap between the metacognitive knowledge elaborated as PO-MK and CO-MK and science teachers' professional knowledge. Much of the literature on science teacher knowledge has been couched in and highlights the importance of teacher reflection. However, as suggested in Figures 3.1 and 3.3, reflection is not metacognition, but it is an important process in developing and enhancing metacognition. What is apparent is that until recently, there has been little, if any, framing of studies into general teacher thinking or teacher knowledge using metacognition theory as a lens through which such thinking might be conceptualized and explored. This is the case even when studies have explored the development of students' metacognition via teacher-led interventions. Only in recent scholarship have Lin et al. (2005), Hall (2006), and Duffy, Miller, Parsons, and Meloth (2009) sought to frame teacher knowledge, thinking, and decision making in terms of metacognition theory. Like much of the "teacher thinking and knowledge" literature, these authors seem more interested in teachers exploring their own strategic and reflective thinking in a more general, "teacherly" sense rather than in terms of subject-specific teacher metacognition. Duffy et al. draw attention to the multiplicity of ways teachers' "conscious, mindful action" (p. 241) might be construed as metacognitive thought; but they argue on the basis of their review of literature that "our question about whether teachers are metacognitive professionals cannot be definitively answered" (p. 247). They go on to propose that research into teacher metacognition is in its infancy, and that "it is assumed that teachers are metacognitive, but more data is needed to document the extent to which they are metacognitive, the factors influencing it, and the effect on students" (p. 247).

Research Opportunities

Teacher metacognition is a potentially fertile field of research. There are very few studies specifically attending to science teacher metacognition as it has been conceptualized above. It is very underresearched, especially as it relates to teachers' metacognitive knowledge and how it can be developed, and also as it relates to teachers' explicit modeling of learning strategies and cognitive processes to learners. It appears this area has fallen, so to speak, "through the cracks." Zohar and Schwartzer (2005) suggest, "we still know relatively little about teachers' knowledge in the area of teaching thinking" (p. 1596). The limited research available paints a fairly bleak picture of research on teacher metacognition in science education. Irrespective of how we conceptualize science teacher thinking and action, it is obvious that considering the metacognition related to that thinking and action enables us to look at teacher thinking through a different lens. Studies that use metacognition frameworks are needed to explore the nature and extent of science teachers' metacognition and the current use of words like reflection, thoughtfulness, and adaptive thinking need to be held up against conceptualizations of metacognition. Any research should take into account the subject material that teachers teach as that which influences the thinking required to understand it. For example, there are similarities and differences in how one learns and thinks about physics, chemistry, and biology; therefore, the metacognition will vary. Finally, it should be long(er) term and employ a range of methodologies. It is not easy to envisage that studies over a day or a week will, for example, be viewed as being particularly useful. Acontextual, clinical studies will have little authority in teacher and science education communities with their concern for context.

CONCLUDING COMMENTS: THE STATE OF SCIENCE EDUCATION AND THE NEED FOR CHANGE

A key objective of science education reform continues to be the development of students' cognitive processes and strategic thinking, especially as it relates to conceptual development, scientific inquiry, and problem solving. Since the early 1960s, there have been massive curriculum reforms in science education, and these reforms have continued unabated. Some science educators, 50 years on, question how much progress has made in improving students' science learning. This skepticism has arisen, at least in part, because reforms in science education consistently ignore issues related to the development and enhancement of students' metacognition. This is despite compelling evidence from empirical research on the importance of metacognition for improving science learning. Thomas (2009b) has previously argued that the importance of metacognition in science education has been largely ignored because science education as a field has its own hegemonic structures and processes, which reproduce the status quo in terms of what is valued in science education; for example, scientific literacy, understanding scientific inquiry, and the nature of science. Reform agendas have typically proposed simplistic solutions to complex problems and tried to routinize and direct the nature of teachers' work. Much of this reform has reflected a government-driven, managerial, top-down approach to change. Teachers are crucial factors in any educational enterprise within schools (Darling Hammond, 2000), yet their thinking and learning processes are typically ignored in reform agendas. Instead, mandates for reform are handed to teachers to implement, and the same unsatisfactory results ensue.

Ideally, teachers' metacognition should become a focus of reform agendas in science education, and across education. Science teachers should be expert science learners as well as science knowers, and knowledgeable about how best to learn science and science concepts. They should be appropriately metacognitive in relation to the cognitive processes by which they and their students learn science. They should be "learning" role models for students, and model and explain learning strategies for students to help take student "apprentices" from novice science learners to more and increasingly competent science learners. They should share and elicit metacognitive experiences with their students, and develop and enhance the metacognitive orientation of science classrooms.

These are substantial expectations that will require several conditions to be met for them to become reality. Firstly, research into science teacher metacognition is urgently needed to build a conceptual basis to inform reform. Secondly, teachers' as well as students' metacognition needs to be given more prominence in teacher education and teacher development. Most educational psychology and science teacher education and in-service courses do not sufficiently investigate metacognition and/or the cognitive processes/strategies required to learn science. Implementing such change is a great challenge. Finally, science teacher educators themselves should be highly metacognitive in all three of aforementioned domains so that they can model the processes of teacher thinking for their prospective would-be science teachers. It is unlikely that these conditions will be universally agreed-upon and met in the short term, but it is hoped that this chapter has at least helped raise awareness of this important issue for the improvement of teaching and learning in science education.

ACKNOWLEDGEMENT

The research for this chapter was supported in part through the Using Metaphor to Develop Metacognition in Relation to Scientific Inquiry in High School Science Laboratories project funded by the Social Science and Humanities Research Council (Canada). Contract grant sponsor: Social Science and Humanities Research Council (Canada). Contract grant number: SSHRC File # 410-2008-2442.

REFERENCES

Abd-El-Khalick, F., & Akerson, V. (2009). The influence of metacognitive training on preservice elementary teachers' conceptions of nature of science. *International Journal of Science Education, 31*(16), 2161–2184.

Anderson, D., Thomas, G. P., & Nashon, S. M. (2009). Social barriers to meaningful engagement in biology field trip group work. *Science Education, 93*(3), 511–534.

Baird, J. R. (1986). Improving learning through enhanced metacognition: A classroom study. *European Journal of Science Education, 8*(3), 263–282.

Berry, A., Loughran, J., Smith, K., & Lindsay, S. (2009). Capturing and enhancing science teachers' professional knowledge. *Journal of Research in Science Teaching, 39,* 575–594.

Blank, L. M. (2000). A metacognitive learning cycle: A better warranty for student understanding? *Science Education, 84*(4), 486–506.

Boekaerts, M. (1997). Self-regulated learning: A new concept embraced by researchers, policy makers, educators, teachers and students. *Learning and Instruction, 7*(2), 161–186.

Borkowski, J., Carr, M., & Pressely, M. (1987). "Spontaneous" strategy use: Perspectives from metacognitive theory. *Intelligence, 11,* 61–75.

Brown, A. L. (1978). Knowing when, where, and how to remember: A problem of metacognition. In R. Glaser (Ed.), *Advances in instructional psychology* (Vol. 2, pp. 77–165). Hillsdale, NJ: Erlbaum.

Chu, M-W. (2009). Exploring science curriculum emphases in relation to the Alberta Physics Program of Studies. Unpublished PhD thesis. Edmonton: University of Alberta.

Connor, L. N. (2007). Cueing metacognition to improve researching and essay writing in a final year biology class. *Research in Science Education, 37*(1), 1–16.

Darling Hammond, L. (2000). Teacher quality and student achievement: A review of state policy evidence. *Education Policy Analysis Archives, 8*(1), 1–50

Davidowitz, B., & Rollnick, M. (2003). Enabling metacognition in the laboratory: A case study of four second year university chemistry students. *Research in Science Education, 33*(1), 43–69.

Dinsmore, D. L., Alexander, P. A., & Loughlin, S. M. (2008). Focusing the conceptual lens on metacognition, self-regulation, self-regulated learning. *Educational Psychology Review, 20,* 391–409.

Donovan, M. S., & Bransford, J. D. (2005). Introduction. In M. S. Donovan & J. D. Bransford (Eds.), *How students learn: History, mathematics and science in the classroom* (pp. 1–27). Washington DC: The National Academies Press.

Duffy, G. G., Miller, S., Parsons, S., & Meloth, M. (2009). Teachers as metacognitive professionals. In D. J. Hacker, J. Dunlosky, & A. C. Graesser (Eds.), *Handbook of metacognition in education* (pp. 241–256). New York: Routledge.

Feuerstein, R., Rand, Y., Hoffman, M., & Millar, R. (1980). *Instrumental enrichment.* Baltimore: University Park Press.

Flavell, J. H. (1971). First discussant's comments: What is memory development the development of? *Human Development, 14,* 272–278.

Flavell, J. H. (1976). Metacognitive aspects of problem solving. In L. B. Resnick (Ed.), *The nature of intelligence* (pp. 231–235). Hillsdale, NJ: Lawrence Erlbaum and Associates.

Flavell, J. H. (1979). Metacognition and cognitive monitoring. *American Psychologist, 34,* 906–911.

Garner, R., & Alexander, P. (1989). Metacognition: Answered and unanswered questions. *Educational Psychologist, 24*(2), 143–158.

Georghiades, P. (2004). From the general to the situated: Three decades of metacognition. *International Journal of Science Education, 26*(3), 365–383.

Georghiades, P. (2006). The role of metacognitive activities in the contextual use of primary pupils' conceptions of science. *Research in Science Education, 36*(1–2), 29–49.

Gess-Newsome, J., & Lederman, N. G. (Eds.). (1999). *Examining pedagogical content knowledge.* Dordrecht, The Netherlands: Kluwer.

Gunstone, R. F. (1994). The importance of specific science content in the enhancement of metacognition. In P. Fensham, R. F. Gunstone, & R. T. White (Eds.), *The content of science: A constructivist approach to its learning and teaching* (pp. 131–146). London: Falmer Press.

Hacker, D. J. (1998). Definitions and empirical foundations. In D. J. Hacker, J. Dunlosky, & A. C. Grasser (Eds.), *Metacognition in educational theory and practice* (pp. 1–24). Mahwah, NJ: Erlbaum.

Hall, E. (2006, July). *Teachers and metacognition: Drawing together evidence from systematic review and action research.* Paper presented at the second meeting of the Metacognition SIG of the European Association for Research on Learning and Instruction, Cambridge, UK.

Hepburn, G., & Gaskell, P. J. (1998). Teaching a new science and technology course: A sociocultural perspective. *Journal of Research in Science Teaching, 35*(7), 777–789.

Kozulin, A. (2005, April). *Who needs metacognition more: Students or teachers?* Paper presented at the annual meeting of the American Educational Research Association, Montreal, Canada.

Lajoie, S. P. (2008). Metacognition, self-regulation and self-regulated learning: A rose by any other name? *Educational Psychology Review, 20,* 469–475.

Leou, M., Abder, P., Riordan, M., & Zoller, U. (2006). Using "HOCS-centered learning" as a pathway to promote science teachers' metacognitive development. *Research in Science Education, 36*(1–2), 69–84.

Lin, X., Schwartz, D. L., & Hatano, G. (2005). Toward teachers' adaptive metacognition. *Educational Psychologist, 40*(4), 245–255.

Loughran, J. J., Berry, A., & Mulhall, P. (2006). *Understanding and developing science teachers pedagogical content knowledge.* Dordrecht, The Netherlands: Sense Publishers.

Macdonald, I. D. H. (1990). *Student awareness of learning.* Unpublished Master of Educational Studies Project. Melbourne, Australia: Monash University.

Mason, L., Boldrin, A., & Ariasi, N. (2010). Epistemic metacognition in context: Evaluating and learning online information. *Metacognition and Learning, 5*(2), 67–90.

Mintzes, J. J., & Novak, J. D. (2005). Assessing science understanding: The epistemological Vee diagram. In J. J. Mintzes, J. H. Wandersee, & J. D. Novak (Eds.), *Assessing science understanding: A human constructivist view* (pp. 41–69). Burlington, MA: Elsevier Academic Press.

Nelson, T. O., & Narens, L. (1994). The role of metacognition in problem solving. In J. Metcalfe & A. Shiminura (Eds.), *Metacognition* (pp. 207–226). Cambridge, MA: MIT Press.

Perkins, D. N. (1992). *Smart schools.* New York: Free Press.

Roberts, D. A. (1982). Developing the concept of "curriculum emphases" in science education. *Science Education, 66*(2), 243–260.

Roberts, D. A. (1988). What counts as science education? In P. Fensham (Ed.), *Development and dilemma in science education* (pp. 27–55). Philadelphia: Falmer Press.

Schraw, G. (1998). Promoting general metacognitive awareness. *Instructional Science, 26*, 113–125.

Sternberg, R. J. (1998). Metacognition, abilities, and developing expertise: What makes an expert student? *Instructional Science, 26*(1–2), 127–140.

Thomas, G. P. (1999). *Developing metacognition and cognitive strategies through the use of metaphor in a year 11 chemistry classroom.* Unpublished PhD thesis. Brisbane, Australia: Queensland University of Technology.

Thomas, G. P. (2002). The social mediation of metacognition. In D. McInerny & S. Van Etten (Eds.), *Sociocultural influences on motivation and learning: Vol. 2. Research on sociocultural influences on motivation and learning* (pp. 225–247). Greenwich, CT: Information Age Publishing.

Thomas, G. P. (2003). Conceptualisation, development and validation of an instrument for evaluating the metacognitive orientation of science classroom learning environments: The Metacognitive Orientation Learning Environment Scale–Science (MOLES-S). *Learning Environments Research, 6*(3), 175–197.

Thomas, G. P. (2004). Dimensionality and construct validity of an instrument designed to measure the metacognitive orientation of science classroom learning environments. *Journal of Applied Measurement, 5*(4), 367–384.

Thomas, G. P. (2006). An investigation of the metacognitive orientation of Confucian-heritage culture and non-Confucian heritage culture science classroom learning environments in Hong Kong. *Research in Science Education, 36*(1–2), 85–109.

Thomas, G. P. (2009a). The centrality of metacognition for science education reform: Challenging the status quo. In A. Kuroda, P. Tanunchaibutra, D. Yaampan, O. Namwong, S. Bhiasiri, & J. Thongpai (Eds.), *Learning communities for sustainable development* (pp. 17–30). Khon Kaen, Thailand: Anna Offset.

Thomas, G. P. (2009b). Metacognition or not: Confronting hegemonies. In I. M. Saleh & M. S. Khine (Eds.), *Fostering scientific habits of mind: Pedagogical knowledge and best practices in science education* (pp. 83–106). Rotterdam, The Netherlands: Sense Publishers.

Thomas, G. P. (2010a, July). *The interview as a metacognitive experience: Insights from research with students.* Paper presented at the 2010 International Conference on Learning, Hong Kong.

Thomas, G. P. (2010b, May). *Changing the metacognitive orientation of a classroom environment to enhance metacognition regarding physics learning.* Paper presented at the fourth meeting of the Metacognition SIG of the European Association for Research on Learning and Instruction, Muenster, Germany.

Thomas, G. P. (in press). Metacognition in science education: Past, present and future considerations. In B. J. Fraser, K. G. Tobin, & C. J. McRobbie (Eds.), *Second international handbook of science education.* Dordrecht, The Netherlands: Springer.

Thomas, G. P., Anderson, D., Ellenbogen, K., & Cohn, S. (2009, August). *Parents' views of learning: Influences on parent-child interactions in a science museum setting.* Paper presented at the conference of the European Association for Research on Learning and Instruction, Amsterdam.

Thomas, G. P., & Au, D. K-M. (2005). Changing the learning environment to enhance students' metacognition in Hong Kong primary school classrooms. *Learning Environments Research, 8*(3), 221–243.

Thomas, G. P., & McRobbie, C. J. (1999). Using metaphors to probe students' conceptions of learning. *International Journal of Science Education, 21*(6), 667–685.

Thomas, G. P., & McRobbie, C. J. (2001). Using a metaphor for learning to improve students' metacognition in the chemistry classroom. *Journal of Research in Science Teaching, 38*(2), 222–259.

Tishman, S., & Perkins, D. N. (1997). The language of thinking. *Phi Delta Kappan, 78*(5), 368–374.

van Driel, J. H., Verloop, N., & de Vos, W. (1998). Developing science teachers' pedagogical content knowledge. *Journal of Research in Science Teaching, 35*(6), 673–695.

Veenman, M. V. J., Van Hout Wolters, B. H. A. M., & Afflerbach, P. (2006). Metacognition and learning: Conceptual and methodological considerations. *Metacognition and Learning, 1*(1), 3–14.

White, B., & Frederiksen, J. (1998). Inquiry, modeling, and metacognition: Making science accessible to all students. *Cognition and Instruction, 16*(1), 3–118.

White, R. T. (1998). Decisions and problems in research on metacognition. In B. J. Fraser & K. G. Tobin (Eds.), *International handbook of science education* (pp. 1207–1213). Dordrecht, The Netherlands: Kluwer.

White, R.T., & Gunstone, R. F. (1992). *Probing understanding.* London: Falmer Press.

White, R. T., & Mitchell, I. J. (1994). Metacognition and the quality of learning. *Studies in Science Education, 23,* 21–37.

Yürük, N. (2005). *An analysis of the nature of students' metaconceptual processes and the effectiveness of metaconceptual teaching practices on students' conceptual understanding of forces and motion.* Unpublished doctoral dissertation, Ohio State University, Columbus.

Zohar, A. (1999). Teachers' metacognitive knowledge and the instruction of higher order thinking. *Teaching and Teacher Education, 15,* 413–429.

Zohar, A. (2004). *Higher order thinking in science classrooms: Students' learning and teachers' professional development.* Dordrecht, The Netherlands: Kluwer Academic Publishers.

Zohar, A., & Nemet, F. (2002). Fostering students' knowledge and argumentation skills through dilemmas in human genetics. *Journal of Research in Science Teaching, 39,* 35–62.

Zohar, A., & Schwartzer, N. (2005). Assessing teachers' pedagogical knowledge in the context of teaching higher order thinking. *International Journal of Science Education, 27*(13), 1595–1620.

Zohar, A., Vaaknin, E., & Degani, A. (2001). Teachers' beliefs about low achieving students and higher order thinking. *Teaching and Teachers' Education, 17,* 465–485.

CHAPTER 4

TEACHING ABOUT CLIMATE CHANGE

An Action Research Approach

John Wilkinson

ABSTRACT

Climate change is a pressing global issue of considerable interest to professional scientists as well as policymakers and even average citizens. It spans
multiple scientific disciplines, and much of the professional science requires
sophisticated and expensive computer models. At the same time, there is a
growing awareness of the need for laboratory experiences in the science classroom that teach both content and science as a method of inquiry. Further
complicating matters are the cost of laboratory equipment and the inherent
variability of climate change impact. However, it is generally accepted that
sea levels will rise due to thermal expansion and glacier melting, so any student living in a coastal area could potentially research climate change. This
chapter reviews the scientific basis of climate change and its potential impact
on society at large. It then proposes a combination of field and laboratory
experience for students who will assess the potential impact of climate change
in their region through the lens of botany. Students or a teacher will collect a
seed bank from the nearest estuary or freshwater marsh. They will then treat

Contemporary Science Teaching Approaches, pages 55–71

the seeds with varying concentrations of saltwater and record the differences in germinations observed. Based on these findings, they will then predict the impact that increased salinity will have on their local biota.

INTRODUCTION

Climate change is an issue of international concern that is expected to increase in importance during the lifetime of today's science students (IPCC, 2007). It could potentially be used as a pedagogical tool for introducing complexity science to students as well as explore the interface of science and public policy (Dahlberg, 2001). However, it can be problematic to teach about climate change in the traditional science classroom (Lenzen & Smith, 2000). It involves multiple sciences, so it is difficult to determine its place in a traditional curriculum. The expense and expertise involved in creating computer-based climate change models makes it difficult to introduce to younger students. However, most models show a clear trend toward increased saltwater intrusion in brackish and freshwater areas (Rapalgia, Athanasios, Bokuniewicz, & Pick, 2010). Therefore, any student living near a coast could potentially engage in action research predicting the potential consequences of climate change on their local flora. This supports the larger trend in education toward classroom research as an instructional pedagogy in all disciplines (Baumfield, Hall, & Wall, 2008).

Background on Climate Change

Hurricane Katrina made landfall near New Orleans on August 29, 2005. It was the costliest hurricane ever, causing damages of about $134 billion (Wright & Boorse, 2010). For many people in the general public, it served as an alarm that human changes to the atmosphere might be causing more intense hurricanes. Indeed, in the 15 years preceding Katrina, every ocean basin showed a statistically significant increase in the number of strong hurricanes (World Resources Institute, 2006). This section will review the science behind climate change, such as increased frequency and intensity of hurricanes as well as some responses to it.

The Earth's atmosphere consists of several spheres. The troposphere is the portion closet to the Earth's surface and is the source of all of the weather and climate experienced by humans on the ground (Miller & Spoolman, 2009). Because it mixes vertically, gases released as pollutants on the surface are quickly mixed throughout the troposphere, often changing chemically or falling in combination with precipitation. Above the troposphere is the stratosphere, which holds ozone that has been depleted by human activities. Beyond that are the mesosphere and thermosphere, which are not pertinent to climate change (Wright & Boorse, 2010).

Solar energy entering Earth's atmosphere can take two paths: Approximately 29% of all solar energy is immediately reflected by clouds or the surface of the Earth. The other 71% is absorbed and re-radiated (Lindsey, 2009). This is the energy that becomes our weather, powers photosynthesis, and therefore makes life as we know it possible. When the absorbed energy is re-radiated, it has changed wavelength to the infrared. Certain molecules, such as carbon dioxide, water, and methane, are able to catch this heat, re-radiate it, and thereby cause increased temperatures near the surface of the Earth. This process is analogous to the glass that holds heat in a greenhouse, hence the popular name greenhouse gases (Miller & Spoolman, 2009).

The existence of a greenhouse effect from certain gases was first recognized by Jean-Baptiste Fourier in 1827 and is noncontroversial. Without any greenhouse gases, the Earth's average temperature would be approximately negative 19 degrees Celsius instead of 14 degrees Celsius (Wright & Boorse, 2010) and as such, the Earth would be much more similar to Mars, and life as we know it would not be possible.

While greenhouse gases have a positive forcing effect on global temperatures, there are a number of negative forcing factors that cool the climate. Some surfaces, such as ice or snow, reflect sunlight, contributing to planetary albedo. This effect is lessened by black soot or melting, creating a positive feedback loop for global warming. Other cooling processes include low-lying clouds, volcanoes, aerosols, and the sun's variability. Since global temperatures depend on the relative impact of a wide variety of natural and human-caused factors, it is difficult to link single weather events to a single cause. However, there is significant evidence that the climate system as a whole has changed to become warmer (IPCC, 2007)

The day-to-day fluctuations in temperature, pressure, wind, and precipitation make up the weather. The long-term aggregate of weather patterns in a particular area can be considered climate. When solar energy is re-radiated from the Earth's surface, it heats the lowest part of the atmosphere, causing it to expand and rise, creating vertical currents. Nearby air rushes in to replace the rising air, creating horizontal currents, or wind. Eventually, the heat in the atmosphere is radiated into space, and it sinks again. This general pattern is responsible for the Hadley cells, and explains the distribution of high rainfall near the equator and deserts between 25 and 35 degrees latitude. Similar smaller-scale convection currents create day-to-day weather patterns and the motion of weather from west to east (Wright & Boorse, 2010).

The interactions of all of these factors create climatic patterns, which then support specific biomes. Plants, animals, fungi, and microscopic organisms have all adapted over geologic time to climatic regimes. While humans are able to adapt to almost all environments on Earth, it may be more difficult for most other living things. Plants might not be able to move or spread seeds quickly enough to adapt to a changing climate. Predator and prey relationships might come unhinged as a result of changes in migrations, hatchings, or other lifecycle events (Miller & Spoolman, 2009).

The consensus of the 2007 Intergovernmental Panel on Climate Change (IPCC) report is that warming of the climate system is unequivocal. This is based on measurements of sea temperatures, atmospheric temperatures, precipitation, glacier melting, extreme weather events, and sea level rise. More specifically, sea temperature measurements indicate that over 90% of the heat increase at the Earth's surface has been absorbed by the oceans (Hansen, 2005). The short-term consequence is that sea level will rise due to thermal expansion. Sea level is currently rising at a rate of about 3.3 millimeters per year (Wright & Boorse, 2010), a phenomena directly explored by the learning activity presented here. Historically, sea level has been very stable, implying a new phenomena consistent with anthropogenic climate change. The longer-term consequence is that this stored heat will eventually come to equilibrium with the atmosphere, meaning that the full consequences of current levels of climate change have not yet been felt, regardless of future mitigation.

There is a wide variety of other observed changes in the global climate consistent with greenhouse gas-induced warming. There has been an overall decrease in extreme cold events and an increase in heat waves. In the northern hemisphere, spring arrives earlier and fall later, sometimes disrupting ecosystems. The amount of land area experiencing drought has doubled in the last 30 years (IPCC, 2007). Arctic sea ice has decreased 11.7% in the last decade, and spring comes an average of two weeks earlier. Both the Greenland and West Antarctic ice sheets are melting at unprecedented rates, threatening to contribute to sea level rise. Precipitation patterns have changed, with a decrease between 10 and 30 degrees north latitude, where a significant portion of humanity lives. Oceans have absorbed about 40% of the atmospheric carbon released, which has begun to acidify oceanic waters. At some unknown level, this will make oceans inhospitable to corals, causing ecological collapse (Miller & Spoolman, 2009).

The IPCC (2007) report goes on to state with over 90% confidence that the warming is caused by human-released greenhouse gasses, such as carbon dioxide, water vapor, methane, nitrous oxide, and halocarbons, such as CFCs. Atmospheric carbon dioxide has risen from 280 to 388 parts per million in the last 150 years. The two major sources for carbon dioxide are the burning of fossil fuels and burning of forests. Currently, 8 billion tons of fossil fuels are burned each year, with almost half coming from industrialized countries (Wright & Boorse, 2010). Global use of fossil fuels continues to rise and is expected to accelerate with the continued expansion of the global economy, particularly in India and China. Burning of forest trees adds an additional 1.6 billion tons of carbon, but also prevents photosynthesis that was acting as a carbon sink. Despite this decrease in forest cover, approximately half of our annual carbon emissions are removed by carbon sinks, such as ocean absorption and photosynthesis. Efforts are underway to mitigate greenhouse gas emissions, as well as adapt to a climate that is already changing (IPCC, 2007).

Water vapor is actually the most abundant greenhouse gas, but its concentration is variable due to constant evaporation and precipitation. A general increase in water vapor has been documented since 1988. As global temperatures increase, more water is expected to evaporate into the atmosphere, creating a positive feedback loop amplifying preexisting global warming (Wright & Boorse, 2010).

Methane is released in much smaller quantities than carbon dioxide, but is 20 times more effective at re-radiating solar energy. Human activities have increased atmospheric methane from historic levels of 714 parts per billion to 1,875 parts per billion (IPCC, 2007). Common sources of methane include ruminant livestock, landfills, coal mines, natural gas production, and rice paddies.

Nitrogen oxide levels have increased 18% over the last 200 years (Wright & Boorse, 2010). It is produced when fossil fuels are burned, particularly gasoline. It is also produced in microbial denitrification processes whenever high levels of nitrogen are available in the soil. Global increase of synthetic fertilizers has caused increased dentirification. Nitrogen oxide lasts an average of 114 years in the atmosphere. This stability allows it to rise to the stratosphere, where it contributes to ozone depletion.

Halocarbons, including CFCs, are human-synthesized chemicals used primarily for refrigeration and other industrial processes. The current atmospheric levels of halocarbons are approximately 1.3 parts per billion, all directly attributable to human activity (IPCC, 2007). These chemicals have also been implicated in ozone depletion. As a result of the Montreal Accord of 1987 to address ozone depletion, CFC levels have begun a slow decline. However, all of the other anthropogenic greenhouse gases are thought to contribute almost as much infrared radiation as carbon dioxide (Wright & Boorse, 2010).

In addition to direct temperature observations, climate scientists often use proxies, such as tree rings, pollen deposits, marine sediments, and ice cores to determine longer-term climate patterns. Analysis of ice cores from Greenland and Antarctica suggest that the Earth's climate oscillates between ice ages and shorter interglacial periods such as our current climatic regime (Miller & Spoolman, 2009). The most likely explanation for these oscillations is the variation in the Earth's orbit over Milankovitch cycles, known periodic intervals taking tens of thousands of years. However, in the ice-core data for the last 800,000 years, carbon dioxide levels oscillated between 150 to 280 parts per million, significantly less than our current level of 388 parts per million (IPCC, 2007).

Beyond the slow changes explained by the Milankovitch cycles, there are instances of rapid change. Studies of the Younger Dryas event about 11,700 years ago suggest that Arctic temperatures rose 7 degrees Celsius in 50 years (Wright & Boorse, 2010). The consensus explanation for these events is disruption in the link between the atmosphere and the oceans. Because oceans have a much greater heat capacity than the atmosphere, they are able to absorb solar radiation, then distribute it globally through thermohaline circulation. Warm, shallow water tends to move to the north Atlantic due to the Gulf Stream, while cold deep and salty water replaces it, moving in the opposite direction. Rapid shifts in climate in the geologic past are thought to result from a disruption of this circulation. Such shifts could occur from the rapid introduction of fresh cold water from glaciers melting into the North Atlantic (Broecker, 2003). The U.S. Climate Change Science Program has concluded that the efficacy of the Atlantic Gulf Stream heat conveyor will decrease gradually over the next century due to freshwater melt from glaciers on Greenland, but will be more than offset by the overall warming caused by increases in greenhouse gases (CCSP, 2008).

Future Projections

Based on computer-based climate models, the IPCC (2007) has predicted two major impacts: regional climate change and rising sea level. For the next century, the extent of both depends on the amount of greenhouse gases emitted worldwide. In the most optimistic scenario, involving immediate global adoption of sustainable technologies that do not rely on fossil fuels, carbon dioxide levels are estimated to level out at 550 parts per million, approximately double pre-industrial levels. This implies a sustained climate change of 2.5 degrees Celsius, which many observers believe will be reached by 2050 (Wright & Boorse, 2010). The predicted impacts of this best-case scenario would be about 15% of all ecosystems transformed. This includes loss of the majority of the remaining Amazon rainforest, global coral reef bleaching, significant melting in polar regions, and an overall extinction of 15% of all species (Kerr, 2007). Because polar regions will continue to experience greater increases in average temperature, over 90% of all permafrost is expected to thaw. This will release large amounts of methane from anaerobic decay, producing another positive feedback loop.

Sea level is currently rising at a rate of 3.3 mm per year, which means it is expected to rise by a total of 190 mm to 590 mm by the end of this century. Due to the slowness of thermal expansion, it is expected to continue to rise for the next two centuries, evening out at between 300 mm and 800 mm above historical levels (IPCC, 2007). There is significant uncertainty about sea level rise, but even a 500 mm rise would flood many coastal areas and increase the impact of flooding and extreme storm events, such as hurricanes. Particularly at risk are low-lying areas that receive minimal benefits from the burning of fossilized carbon, such as large areas of Bangladesh and many Pacific island nations.

One limitation is that these estimates do not consider any increase in melting from land-based glaciers in polar regions. An increase of 3 degrees Celsius will create conditions that could cause melting of most of Greenland's ice sheet. In Antarctica, 87% of the glaciers have shown a net retreat in the last 50 years. Satellite measurements suggest that the Antarctic ice sheet is currently losing enough water to raise sea level by an additional 0.4 mm a year. Models that include the plausible simultaneous melting of Greenland and Antarctica's glaciers suggest a sea level rise of between 800 mm and 2,000 mm (Wright & Boorse, 2010). Even if greenhouse gas levels are stabilized, sea levels will continue to rise for hundreds of years, causing hardships for the half of the world's population that live in coastal regions.

With increased freshwater at the ocean's surface, the North Atlantic thermohaline circulation is expected to slow, but not reverse. As more-intense storms occur, extreme wave heights and precipitation events will increase. However, many already dry regions will receive less rainfall. Approximately 30% of the Earth's land surface will experience droughts, while other regions will experience increased flooding.

Mitigation

While some adaptation will be necessary in the immediate future, mitigation is the only long-term solution. The arguments for rapid mitigation include environmental ethics, such as the polluter pays and the precautionary principle. The polluter pays principle of most environmental legislation is difficult to implement for global warming because the costs of climate change are separated in both space and time from the benefits. A universal carbon tax or cap and trade scheme might offer some means to link costs with benefits. The 1992 Rio Earth Summit Declaration states, "Where there are threats of serious or irreversible damage, lack of full scientific certainty shall not be used as a reason for postponing cost-effective measures to prevent environmental degradation" (UNEP, 1992). For climate change, this means lowering fossilized carbon emissions to a level that allows for a margin of error given current uncertainties in atmospheric science. There is also a very strong economic argument for immediate mitigation. The British Stern Review estimates that inaction on climate change would result in a 5%–20% decrease in global GDP per year, for the indefinite future (Stern, 2007).

Successful mitigation requires that the carbon dioxide level does not dangerously interfere with the climate system. An increase of 3 degrees Celsius is enough to raise sea levels 25 m. Two degrees Celsius could melt the Greenland Ice Sheet, so 1 to 1.5 degrees Celsius seems to be the most prudent level to assure a climate similar to the one in which our agricultural system evolved. This would require stabilization of carbon dioxide levels at approximately 450 parts per million, which means cutting carbon emissions by 75% by the end of this century. The IPCC (2007) analyzed six possible carbon stabilizing scenarios, and only the lowest one would possibly limit global warming to a 1.5 degree Celsius rise in temperature. They go on to urge immediate mitigation efforts since the next decades will have a large impact on final carbon stabilization levels (Wright & Boorse, 2010).

Previous attempts at mitigation have met with mixed success. The 1992 Framework on Climate Change signed in Rio de Janeiro adopted a voluntary reduction approach (UNEP, 1992). However, since the framework was voluntary, almost all the countries that signed ignored their pledge and continued to increase their carbon emissions. The Kyoto Protocol went into effect in 2005 with 189 countries participating. It requires all participating developed economies to lower their emissions of carbon dioxide, methane, and nitrous oxide to lower than 5.2% of their 1990 emissions. The Protocol allows for caps on emissions and trading within and between countries. In 2001, then-president George Bush withdrew the United States from the Kyoto Protocol, citing its negative effects on the U.S. economy and the lack of emissions requirements for developing economies, such as China or Brazil. Advocates for the protocol hope that these economies will agree to reduce their emissions during the second phase, which begins in 2012 (Miller & Spoolman, 2009).

In late 2009, world leaders met in Copenhagen, Denmark, in an attempt to create a treaty that would be more effective in mitigating climate change than the Kyoto agreement. Instead, only a much more modest deal between the United States and four countries was acknowledged. This happened primarily due to the political interests of the Obama administration as well as politics in China and India.

There are many options for future mitigation of greenhouse gas emissions. There are three major means for creating a financial signal to mitigate fossil fuel use: removing subsidies for fossil fuels, creating a carbon tax, or creating a market-based cap and trade system. The removal of subsidies would be the most straightforward approach, given the fiscal difficulties of many governements. Governments worldwide provide depletion allowances, tax relief to consumers, support for exploration, and military security for multinationals extracting in the Middle East. While this might appear to be a clear win-win situation, fossil fuel companies are significant contributors to political campaigns, making any change in the current regime difficult at best.

The carbon tax has the advantage of being spread widely through the economy, including new industries that have not yet been invented. It would also serve as a constant reminder to both producers and consumers of the costs inherent in fossilized carbon emissions. However, since poor people generally spend a greater portion of their income on energy, it would be a regressive tax unless it could be coupled with a decrease in payroll taxes or some similar mechanism (Friedman, 2008).

A cap and trade system would provide a global limit on carbon emissions and permits to different companies that would be tradable. This gives a financial incentive for technological improvement below government mandates and allows for great engineering flexibility. The cap on emissions would be lowered in stages to allow for an eventual stabilization of the atmosphere. It would produce significant revenues to address unforeseen difficulties. However, older industries would be favored, while new industries might not receive sufficient permits to become viable (Wright & Boorse, 2010).

Other ideas for mitigation include increasing the development and distribution of renewable energy sources, such as solar, wind, bio-fuels, hydrogen-powered vehicles, and geothermal. There is the potential for carbon capture and sequestering, which is especially important for coal-powered plants. Nuclear power could also mitigate some carbon emissions, but has issues of cost-effectiveness and waste management. Reforestation would use the natural process of photosynthesis to capture atmospheric carbon as well as potentially improving other environmental issues such as erosion and species extinction. Increasing efficiency of the energy already produced would reduce the need for some carbon emissions (Pacala & Socolow, 2004). The Carbon Mitigation Initiative has developed a game, free to educators, that illustrates the trade-offs needed to flatten the world's carbon emissions (http://cmi.princeton.edu/wedges).

Adaptation to some climate change is inevitable, given the amount of fossilized carbon already in the atmosphere, as well as the time lag between initiation and implementation of a comprehensive public policy on carbon emissions. All countries will have to adapt to climate change to a certain extent. Farmers may have to grow climate-resistant varieties or switch crops entirely. Irrigation might have to be expanded in some areas, while it might be untenable to maintain in other areas. Infrastructure, especially in low-lying coastal areas, will have to be improved. This will include new or improved sea walls, levees, reservoirs, and regrowth of vegetation in wetlands areas. Early-warning systems and evacuation procedures will have to be improved, especially in areas that have not historically experienced extreme weather events. Shelters and resettlement procedures might also need to be developed (Stern, 2007).

While most of the carbon already emitted was burned in developed countries, the IPCC (2007) predicts that in terms of both economic and social costs, developing countries will experience the greatest impact. Investments in traditional human development, such as education, public health, and food security, will be needed to increase adaptive ability in the face of extreme weather. Contagious diseases that are already an issue can be expected to intensify with increased numbers of climate refugees. In addition, incidence of diseases that rely on an animal vector, such as malaria or West Nile Virus, will expand as the vector's range expands. Finally, diversified economies might be necessary to withstand the unknown changes in the future economic climate (Wright & Boorse, 2010).

Impact of Climate Change on Wetlands

Marshes and estuaries are some of the most biologically diverse and productive ecosystems in the world; they provide breeding grounds for fish, shellfish, and birds (Chmura, Anisfeld, Cahoon, & Lynch, 2003). They are also among the most affected by human activity: seaside development, agricultural and industrial pollution, and irrigation projects have all impacted marshes and estuaries generally (Atwater et al., 1979). They are also disproportionately affected by invasive species (Stephens & Sutherland, 1999). Global climate change is expected to increase damage to already fragmented ecosystems (Goldenberg, Christopher, Alberto, & Gray, 2001). This might further increase susceptibility to invasive species (Taylor, Davis, Civille, Grevstad, & Hastings, 2004).

Anthropogenic climate change is expected to contribute to this impact through saltwater intrusion from three different factors (Knowles & Cayan, 2002). First, encroachment of saltwater due to rising sea levels primarily caused by thermal expansion in the short term and the melting of glaciers, particularly those of Antarctica and Greenland in the long term (Rapalgia et al., 2010). Secondly, precipitation that has historically fallen as snow might fall as rain, causing pulses of freshwater followed by significantly less snow-melt through the dry summer months (McCabe & Wolock, 2010). Finally, aging dikes will be exposed to increasing pressure from rising sea levels and storms. If this causes a dike to break, freshwater seed banks would be exposed to a sudden influx of saltwater (Mount & Twiss, 2005). Increases in saltwater due to shifts in precipitation and snowmelt are predicted to have an immediate impact on marshes and estuaries, but sea level rise is a greater long-term threat to tidal marsh vegetation (Knowles & Cayan, 2002).

One clear case of these impacts is the San Francisco Bay. Over 60% of the historical freshwater flow has been diverted for irrigation and municipal use. In dry years, which are expected to become more common due to climate change, up to a third of the remaining flow is diverted. This has allowed for saltwater intrusion from the Pacific Ocean. The results have been devastating both ecologically and economically. Major commercial fisheries in salmon, sturgeon, Dungeness crab, and striped bass have been greatly reduced without financial compensation to the fishermen. The total area covered by wetlands is now only 8% of its historical area, and most of that is home to invasive species (Wright & Boorse, 2010).

Up the San Joaquin and Sacramento Rivers, over a thousand miles of poorly constructed levees protect farmlands that have sunk an average of 20 feet due to compacting from groundwater removal. Since this is a seismically active area, an earthquake could easily rupture the levees, thereby sending a surge of saltwater upriver into the pumps that feed irrigation and municipal sources. In recognizing these problems, the state of California and the U.S. federal government have collaborated to create an integrated management approach to restore ecological health and improve water management. However, the results have been decidedly mixed due to infighting and disagreement between major stakeholders over how to implement recommended changes (Wright & Boorse, 2010).

Previous research indicates that increasing salinity is negatively correlated with increasing biodiversity: if freshwater marshes become saline, plant biodiversity can be expected to decrease, along with its associated fauna (Atwater et al., 1979). Seed banks in particular provide the systemic robustness that maintains current plant community structure (Leck, Parker, & Simpson, 1989). However, many freshwater and brackish wetlands have undocumented seed banks (Leck, Baldwin, Parker, Schile, & Whigham, 2009). The potential therefore exists for students to engage in meaningful primary science research into their local wetlands seed banks as well as model the potential impact of climate change on those seed banks.

The early stages of a plant's life cycle are the most vulnerable to environmental change. Specifically, research by Ayers (1951, cited in Unger, 1962) suggests that most plants will not even germinate in the presence of saltwater. If this is true for the seed bank of a freshwater marsh, the entire plant community would shift vegetation toward the few plants that are salt tolerant (McKee & Mendelsshon, 1989).

Figure 4.1 The author collecting the seed bank of a fresh water estuary near the San Francisco Bay, California. Photo credit: Tom Parker.

Methods

This action research model for high school biology students has two parts: a field experience to collect samples of the local seed bank and a classroom or greenhouse portion to grow the plants. If it is untenable to bring a group of students into a local wetland, the teacher might want to collect the soil samples and focus on the classroom portion. If possible, however, it is advisable to bring students into the field since it will allow them to see the entire scientific process in action, as well as become more familiar with their local environment.

In temperate zones, seed bank collections should be conducted in the winter in order to ensure the seeds will be collected before germination. If the collections occur in winter, they should be refrigerated at 4 degrees Celsius, which should provide stratification.

Field Portion

Equipment required:

1. Transportation into wetland area.
2. Corer, trowel, or shovel to gather soil samples (one for every five students).
3. Plastic bags to hold samples.
4. Markers to note location of each sample.

Each student should be assigned a collection site. While all sites should be chosen at random; individual sites should cover three areas: on the interface between the land and the nearest major body of water, inland, and along any channels in the area under study. If possible, a minimum of 20 meters should separate the students collection site. A total of 10 soil cores (5 cm deep) should be taken randomly within a meter and a half of each site, following Parker and Leck (1985). Each core should be approximately 8 cm in diameter at each plot. As an advanced option, standing vegetation at each site may be recorded.

Lab Portion

Equipment Required:

1. Plastic quarter flats
2. Sand to line all flats
3. Source of ocean or other saltwater
4. Source of freshwater
5. Watering cans (minimum of three)

Once soil samples are collected, they should be brought back to a classroom or greenhouse. The 10 cores from each sampling location should be aggregated by hand, and all coarse organic material removed, including roots and rhizomes. The one homogenous sample from each plot should then be divided equally by weight into samples for three different treatments, following Dailey and Parker (2009). The soil samples should then be transferred to quarter flats, potted with sand, and flushed with freshwater to stimulate germination under greenhouse conditions.

Figure 4.2 The author with a soil core near San Francisco, California. Photo credit: Tom Parker.

The preparation of treatments for the soil samples provides a good opportunity to introduce or review the basic chemistry of solutions. Students should prepare treatments as follows: the first treatment should be a freshwater/control of 0 parts salt per thousand parts water (0 ppt), the second should be one part saltwater to ten parts freshwater, and the third treatment should be all ocean or other saltwater. Students should flush their seed bank samples with the respective solution daily.

Records should be kept and statistically analyzed at a later date for final results. A data chart should be prepared before germinations begin, including space for each species and in which treatment flat the germination occurred. Students should identify or give an unknown name used by the entire class to each species of plant. When no additional germination has been recorded for one week, it should be determined that the treatment is completed, after which statistical analysis of the results should commence.

This statistical analysis should investigate several issues. The first, based on the control results, is to determine the existing seed bank, which might be unknown. The impact of varying concentrations of saltwater on the existing seed bank should also be determined. Gross and species-specific germination rates for the two saline treatments should be compared against the rates of the freshwater/control treatment to examine whether varying degrees of salt inhibited germination and, if so, to what degree. Based on previous research, it is expected that overall germination rates for most species will decrease with increasing salt concentrations (Ayers, 1951, cited in Unger, 1962). However, any halophytic plants are expected to be unaffected by saltwater treatments. Finally, germination rates for site types should be compared to examine whether varying location affected the seed bank. Salt-inhibited germination is expected to be greatest in the channel seed bank, since previous research suggests it will have the greatest species diversity, including those most vulnerable to salt (Leck, Simpson, & Parker, 1989).

Advanced students may be able to compare this seed bank to the existing vegetation. It is expected that the current seed bank will correspond closely with the standing vegetation, with the exception of greater representation of plants that have vegetative reproduction strategies, such as rhizomes, in the standing vegetation since they were outside the scope of this experiment.

CONCLUSION

Global climate change presents a difficult topic to convey to students for a number of reasons: the science is multidisciplinary, so it does not fit neatly into the existing discipline structure; it is complex and relies on sophisticated computer models beyond the scope of secondary science instruction; and the data are constantly being updated, so it is difficult to know how it will impact specific communities. One potential way to convey information about climate change, and to engage students in action research, is to model the impact of saltwater intrusion on local wetland plant communities.

Wetlands are important for biologists for two reasons: They tend to have extremely high rates of primary productivity, and they are among the most affected ecosystems in the world. These affects might be amplified by anthropogenic climate change. Sea level rise due to thermal expansion and land-based glacier melt is expected to cause saltwater intrusion in many of the world's freshwater marshes. This is expected to radically change the composition of the plant communities and lower primary productivity. Plant mortality due to salinity is expected to be greatest in the early life stages, which is where this project is focused.

This chapter outlines one possible procedure to provide both laboratory and field experiences for secondary biology students that will simultaneously develop their knowledge of their local flora, the scientific method, and the impact of climate change in their local environment. In the field component, students learn how to take random samples and identify the standing local flora. In the greenhouse portion, students learn how to create a series of solutions, observe the early life cycle of local freshwater plants, use a control, organize and analyze data systematically, and even potentially publish the results.

REFERENCES

Atwater, B., Conard, G., Dowden, J., Hedel, C., MacDonald, R., & Savage, W. (1979). History, landforms, and vegetation of the estuary's tidal marshes. In T. Conomos (Ed.), *San Francisco bay: The urbanized estuary* (pp. 347–385). San Francisco: Pacific Division AAAS.

Baumfeld, V. Hall, E., & Wall, K. (2008). *Action research in the classroom.* Thousand Oaks, CA: Sage.

Black, R. (2009, December 22). Why did Copenhagen fail to deliver a climate deal? *BBC News.* Retrieved from http://news.bbc.co.uk/2/hi/science/nature/8426835.stm

Broecker, W. (2003). Does the trigger for abrupt change reside in the ocean or in the atmosphere? *Science, 300,* 1519–1522.

Chmura, G., Anisfeld, S., Cahoon, D., & Lynch J. (2003). Global carbon sequestration in tidal, saline wetland soils. *Global Biogeochemical Cycles, 17*(4), 1111.

Climate Change Science Program (CCSP). (2008). *Abrupt climate change.* Reston, VA: USGS.

Dahlberg, S. (2001). Using climate change as a teaching tool. *Canadian Journal of Environmental Education (CJEE), 6*(1), 9–17.

Dailey, B., & Parker, V. (2009). *Effects of salinity on seed bank germination and seed bank viability in a freshwater system.* Unpublished master's thesis, San Francisco State University.

Friedman, T. (2008). *Hot, flat and crowded.* New York: Farrar, Straus & Giroux.

Goldenberg, S., Christopher, W., Alberto, M., & Gray, W. (2001). The recent increase in Atlantic hurricane activity: Causes and implications. *Science, 293,* 474–479.

Hansen, J. (2005). Earth's energy imbalance: Confirmation and implications. *Science, 308,* 1431–1435.

Intergovernmental Panel on Climate Change (IPCC). (2007). *Climate change 2007: The physical science basis.* Contribution of Working Group I to the fourth assessment report of the Intergovernmental Panel on Climate Change. New York: Cambridge University Press.

Kerr, R. (2007). How urgent is climate change? *Science, 318,* 1230–1231.

Knowles, N., & Cayan, D. (2002). Potential effects of global warming on the Sacramento/San Joaquin watershed and the San Francisco estuary. *Geophysical Research Letters, 29,* 1–5.

Leck, M. A., Baldwin, A., Parker, V., Schile, L., & Whigham, D. (2009). Plant communities of tidal freshwater wetlands of the continental USA and Canada. In A. Barendregt, D. Whigham, & A. Baldwin (Eds.), *Tidal freshwater wetlands* (pp. 41–58). Leiden, The Netherlands: Backhuys Publishing.

Leck, M. A., Parker, V. T., & Simpson, R. L. (Eds.). (1989). *Seedling ecology and evolution.* New York: Academic Press.

Leck, M. A., Simpson, R. L., & Parker, V. T. (1989). The seed bank of a freshwater tidal wetland and its relationship to vegetation dynamics. In R. R. Sharitz & J. W. Gibbons (Eds.), *Freshwater wetlands and wildlife* (pp. 189–205). DOE Symp. Series No. 61. CONF-8603101, Oak Ridge, TN: USDOE Office of Sci. and Tech. Information.

Lenzen, M., & Smith, S. (2000). Teaching responsibility for climate change: Three neglected issues. *Australian Journal of Environmental Education, 15,* 65–76.

Lindsey, R. (2009, January 14). Climate and Earth's energy budget. *NASA, Earth Observatory.* Retrieved from http://earthobservatory.nasa.gov/Features/EnergyBalance/

McCabe, G., & Wolock, D. (2010). Long-term variability in northern hemisphere snow cover and associations with warmer winters. *Climatic Change, 99*(1–2), 141–153.

McKee, K., & Mendelsshon, I. (1989). Response of a freshwater marsh plant community to increased salinity and increased water level. *Aquatic Botany, 34*(4), 301–316.

Miller, G., & Spoolman, S. (2009). *Sustaining the Earth: An integrated approach* (9th ed.). Belmont, CA: Brooks/Cole Inc.

Mount, J., & Twiss, R. (2005, March). Subsidence, sea level rise, and seismicity in the Sacramento-San Joaquin delta. *eScholarship, University of California, San Francisco Estuary and Watershed Science, 3*(1), 5. Retrieved from http://repositories.cdlib.org/jmie/sfews/vol3/iss1/art5

Pacala, S., & Socolow, R. (2004). Stabilization wedges: Solving the climate problem for the next 50 years using current technologies. *Science, 305,* 968–972.

Parker, V., & Leck, M. (1985). Relationships of seed banks to plant distribution patterns in a freshwater tidal wetland. *American Journal of Botany, 72,* 161–174.

Rapaglia, J., Athanasios V., Bokuniewicz, H., & Pick, T. (2010). *Forecasting saltwater intrusion into coastal aquifers due to climate change.* Paper presented at the 8th Symposium on Groundwater Hydrology, Quality, and Management, Groundwater Council (pp. 752–760). May 16–20.

Stephens, P., & Sutherland, W. (1999). Consequences of the Allee effect for behavior, ecology and conservation. *Trends in Ecology & Evolution, 14,* 401–405.

Stern, N. (2007). *The economics of climate change: The Stern Review.* Cambridge, UK: Cambridge University Press.

Taylor, C., Davis, H., Civille, J., Grevstad, F., & Hastings, A. (2004). Consequences of an Allee effect in the invasion of a Pacific estuary by *Spartina alterniflora. Ecology, 85,* 3254–3266.

Unger, I., (1962). Influence of salinity on seed germination in succulent halophytes. *Ecology, 43*(4), 763–764.

United Nations Environmental Program (UNEP). (1992). *Rio Declaration on environment and development.* Retrieved from http://www.unep.org/Documents. Multilingual/Default.asp?documentid=78&articleid=1163

World Resources Institute. (2006). *Hot climate, cool commerce: A service sector guide to greenhouse gas management.* Washington, DC: WRI.

Wright, R., & Boorse, D. (2010). *Environmental science: Toward a sustainable future.* New York: Pearson Inc.

John Wilkinson, PhD, Assistant Professor of Education at Bahrain Teacher's College, Kingdom of Bahrain, has been involved in life and environmental science education for 13 years. He taught high school science for 5 years while completing his Master's in Teaching from the University of San Francisco. He then pursued a PhD in Humanities. While completing his dissertation in the philosophy of biology, he lived and taught at Yantai University in China. Upon completion of his PhD, he taught a variety of classes in the Liberal Studies Department of the Art Institute of California, San Francisco. He has most recently been teaching at Bahrain Teacher's College. His current research focuses on wetlands ecology education. In 2008, he won a grant with the Organization of Tropical Studies to research the impact of shrimp farming on mangrove forests in Central America. In 2010, he completed an NSF-funded research project on the impact of climate change on the freshwater marsh communities in the San Joaquin, California river delta.

CHAPTER 5

CONSTRUCTIVIST, ANALOGICAL, AND METACOGNITIVE APPROACHES TO SCIENCE TEACHING AND LEARNING

Samson M. Nashon and J. Douglas Adler
University of British Columbia, Canada

ABSTRACT

Quite often, children have ideas that are different from the scientifically "correct" ones (canonical ideas or concepts). Hence, the aim of teaching is to help children understand and appreciate the "correct" view of science and probably make a shift from the "nonscience" ideas (misconceptions or alternative frameworks) to the "correct" (canonical) ideas. This chapter discusses analogical strategies that embrace key constructivist principles that embody inquiry-framed science pedagogy intended to bring about the conceptual in students' understanding. By employing analogical tools of making sense of new learning, this chapter will demonstrate the links between analogical learning and conceptual change theory. This is because analogical teaching and learning is necessarily constructivist. Constructivist pedagogy is necessarily inquiry based.

Contemporary Science Teaching Approaches, pages 73–113
Copyright © 2012 by Information Age Publishing

73

CONSTRUCTIVISM AND LEARNING

Advocates of constructivism argue that knowledge is constructed as the learner interacts with experience(s) (Driver, 1983, 1989; Driver & Erickson, 1983; Kelly, 1955; Nashon, 2004; Nashon & Anderson, 2004). Understanding and interpretation of new experiences is dependent on the knowledge the learner already has (Ausubel, 1963; Driver, 1983; Hewson & Thorley, 1989; Hodson, 1998; Kelly, 1955; Matthews, 1994; Posner, Strike, Hewson, & Gertzog , 1982).

A prominent approach that describes how the learner interacts with new information is Piaget's model of equilibration (1970). This model explains how learners take in new information from their environment, how they perceive and encode it from outside, and integrate it into their own knowledge (Piaget, 1970). The equilibrium theory describes the way learners try to maintain a balance between the environmental information on the one hand and their prior knowledge on the other (Piaget, 1970). According to Dale (1975), equilibrium means a compensation for an external disturbance. The compensation is by having "a changed viewpoint due to alterations to existing knowledge and modification of existing mental structures to incorporate the new aspects" (pp. 117–118). If information is new and not in line with existing knowledge, this incongruity causes a cognitive conflict (Piaget, 1970). When information cannot be promptly decoded and integrated into existing knowledge, people have to adapt to the new environment. Piaget points out that such cognitive conflicts can lead to new knowledge. There are two possibilities to solve a cognitive conflict: by assimilating the new information or accommodating the knowledge to make it compatible with new information (Piaget, 1970).

Assimilation

Assimilation refers to the process whereby an individual understands new information on the basis of existing knowledge and integrates it into prior knowledge. Assimilation describes the quantitative aspect of individual learning only as additional pieces of information that fit into existing knowledge are added (Piaget, 1970).

Accommodation

The other process of adaptation is the process of accommodation whereby people interact with new information in a way that changes their knowledge. They don't simply assimilate new information into existing

© 1956 United Features Syndicate, Inc.

Figure 5.1   The dual process of assimilation–accommodation.

knowledge, but actually change knowledge in order to better understand the environment and its information. This creation of new knowledge refers to the qualitative manner of learning.

Conceptual Change Model

Most of the models proposed to explain conceptual change have emphasized the role of cognitive conflict as a central condition for conceptual change. Describing the processes of equilibration, Piaget (1970) considered cognitive conflict as a step in this process. The pioneer model of Posner et al. (1982) considered the phase of conflict, generated by dissatisfaction with the existing concepts, as a first step to achieving conceptual change (Limon, 2001). According to Posner et al., the conceptual change model of learning is about the relative status of the existing (idea) and the incoming new information. For a concept to be accepted by the learner, its status must be raised and the old idea lowered in status. In this situation, the new idea is accommodated: the old is replaced by the new, the new idea/concept/information is modified to be consistent with the old, or the old framework (idea) is reorganized to embrace the new. But if the old is as good as the new such that the learner finds both useful, then the new idea is assimilated. In these two scenarios, the learner undergoes conceptual change (Carey, 2000; Duit & Treagust, 2003; Hewson & Thorley, 1989; Posner et al., 1982). What this means is that the learner has added new information to what he/she already had or has changed his/her original viewpoint. Simply put, the learner now sees things differently, hence the idea of conceptual change.

The Predict–Observe–Explain (POE) model (Gunstone, 1994; White & Gunstone, 1992) can be regarded as a practical model for implementing Posner and colleague's (1982) conceptual change framework. These two models—the conceptual change (CC) and POE—are constructivist and predicated on the following tenet assumptions underlying constructivism:

- Knowledge is constructed
- Students already have ideas about natural phenomena
- Interpretation of learning experiences is in part based on knowledge already possessed (Kelly, 1955; Nashon, 2006).
- Effective pedagogy attends to students' prior knowledge by designing activities that elicit and challenge counter physics views.

In the POE model,

1. Students **P**redict the outcome of an event and explain their prediction
2. Students **O**bserve the event
3. Students revise their initial **E**xplanation if inconsistent with their initial prediction

In the conceptual change model, a key condition is that there must be a cognitive dissonance after the event, that is, prior knowledge is inadequate to explain the observation (thus exploiting number 2 of POE). The third step of POE is met by the last three conditions for accommodation, whereby the new idea must be intelligible (sensible), plausible (reasonable), and fruitful (useful). However, not all that is presented to learners is accepted. Some students are resistant to conceptual change because of their various psychological or sociocultural beliefs. Hodson and Hodson (1998a) have interpreted this resistance by viewing such events from a Vygotskian perspective. This is an outlook that sees the group or social class as having a major role in shaping what its members learn or accept as useful information.

Among the factors influencing learning or acceptance of a new concept is social pressure. This may come, say, from other peers who hold a particular view. The peers could be those who occupy a special place in the student's personal life with respect to what Hodson (1998) calls the "student power" hierarchy. They create immense social pressure to conform, leading to a development and legitimization of a change of view. Sometimes the learner may hold membership of several social groups, each of which has rules to follow and exerts social pressures to conform, in Hodson's view, "The personal framework of understanding view of learning allows for the proliferation of meaning in response to entry into additional social groups" (p. 53).

Social and cultural group identities such as gender, ethnicity, religion, and politics also impact heavily on learning (Hatano & Inagaki, 2003; Hodson, 1998). Each of these cultures or subcultures (Aikenhead, 1996; Aikenhead & Ogawa, 2007) provides the learner with different "lenses" through which experiences are interpreted and understood. These different ways of "seeing things" can develop into what Claxton (1993) calls "stances." Stances tend to direct or guide individual students and their group activities or actions. In other words, stances constitute the ethics, the code of practice, or rules that individual members of a group and the group as a whole subscribe to. These stances become the frameworks that guide students' learning behavior. Sticking to these stances is like ensuring allegiance to the norms and practices that constitute them. It is because of this that students always work to hold together all the social groups they associate with. They do this by implicitly or sometimes explicitly swearing allegiance to the social groups. Some of the learners have been found to "act out" by supplying responses that depend on the prevailing circumstances (Hodson, 1998; Matthews, 1994).

The constructivist approach has recently come under serious scrutiny. It is charged by some people with being simplistic, especially in its account of how misconceptions can be eliminated (Bloom, 1995; Cobern, 1996; Hodson & Hodson, 1998a, 1998b; Matthews, 1998). These authors seem to concur on the point that the process of meaning-making or construction is context bound and includes emotions, values, aesthetics, interpretive frameworks, and personal experiences. It is precisely this perspective that Lagoke, Jegede, and Oyebanji (1997) attempted to address by exploring the potential of environmental analogs in enhancing science concept learning among gender groups in Nigerian schools. In other words, they recognize the importance of social factors in analogical learning. This way, analogical learning is of a social constructivist nature. If learners already hold misconceptions in the analog domain (known), then it follows that the same or hybrid misconceptions will be transferred to the target domain (new). In other words, analogies do not necessarily eliminate misconceptions.

Clearly, a construct resulting from analogies should be in line with students' correct preconceptions. Kelly (1955) defines a construct as simply a way in which some things are construed as being alike and yet different from others. Essentially, this is how an analogy works; presenting things or concepts that are different as though they are the same. The analog and target concepts in an analogy are different, and yet the matching of attributes in these two domains or paradigms works as though the matched attributes are alike.

Analogical Teaching and Learning

The meaning of analogy, metaphor, model and example, and how they are related, will be discussed in detail. In addition, Piagetian Schema theories and social constructivist ideas in learning will be discussed as well. Three theoretical models for developing an analogy—GMAT, TWA, and WWA—are also discussed.

THE MEANING OF ANALOGY

Explaining new concepts using similar known and familiar concepts constitutes an analogy. The new concepts are the unfamiliar topics or ideas and the known concepts are the familiar topics or ideas. Therefore, use of familiar ideas, models, situations, and maps to explain the unfamiliar constitutes an analogy (Black & Solomon, 1987; Duit, 1991; Nashon, 2000, 2004; Stepich & Newby, 1988; Thiele & Treagust, 1994; Weller, 1970; Zeitoun, 1984). The familiar concepts (ideas, models, situations, and maps) are called analogs or bases, and the unfamiliar concepts are called targets or topics (Glynn, 1991; Harrison & Treagust, 1993; Lagoke et al., 1997; Nashon, 2000, 2004; Zeitoun, 1984).

The actual analogy involves comparison of features of both the analog and the target. These features are sometimes referred to as attributes (Glynn, 1991; Harrison & Treagust, 1993; Zeitoun, 1984). An analogy is complete if attributes in the analog match those in the corresponding target. Such attributes are said to be similar; it is these similarities that give the analogy its power. However, in practice, not all the attributes in the two domains of the analogy match or are similar.

Stretching further, the range of characteristics of an analogy includes relating abstract ideas to the real world and, in some cases, involves opening new points of view or perspectives (Brown & Clement, 1989; Treagust, Duit, Joslin, & Landauer, 1992). In addition, analogies promote visualization of the abstract and the unfamiliar. In other words, analogies make the invisible and abstract appear visible and real. For example, the familiar water circuit analogy assists mental visualization of the behavior of electric current in a circuit by using the behavior of water in a water circuit. Water represents current. Zeitoun (1984) sums up the meaning of an analogy by saying that when "used as a teaching strategy, [an] analogy provides a comparison, which can explain something difficult to understand by pointing out its similarities to something easy" (p. 107). This view has also been expressed by Kilbourn (2000), who says, " helpful analogies tend to be less complex than the phenomena they are meant to depict" (p. 3). In other words, analogies have to be simple and meaningful.

In a teaching situation, some analogies are planned in advance of instruction while others are spontaneously generated during instruction. Moreover, some of these analogies are universal while others are local. For the purpose of this discussion, universal analogies refer to those found in textbooks and widely understood across different social and cultural contexts, while local analogies apply to those drawn from students' immediate environment. They are understood only within the immediate teaching/ learning context. Lagoke et al. (1997) refer to such analogs as "environmental," in the sense that they are analogs drawn from the sociocultural environment of the learner.

Whether planned or spontaneously generated, universal or local, analogies that teachers use tend to work well. However, there are times when analogies convey to students wrong meanings of the new concepts (targets), as revealed during their contribution to classroom or group discussions, or their responses to test questions on analogous concepts (Gunstone, 1994; Kelly, 1955). Therefore, it is important to have good research data on the effective use of analogies and to identify any commonly used analogies that do not function well. This is because the repertoire of good analogies that teachers and textbooks possess is inadequate (Black & Solomon, 1987; Webb, 1985). According to Willner (1964),

> Science is an immensely creative and enriching experience; and it is full of novelty and exploration; and it is in order to get to these that analogy is an indispensable instrument. Even analysis, even the ability to plan experiments, even the ability to sort things out and pick them apart presupposes a good deal of structure, and that structure is characteristically an analogical one. (p. 479)

To emphasize further how important analogies are, Lagoke et al. (1997) say,

> The use of analogy has been found to be beneficial in science learning by motivating students, providing visualization of abstract concepts, providing a basis for comparing similarities of students' worldviews with new concepts, promoting associations with other experiences, overcoming misconceptions, and coping in the classroom with the complexity of students' beliefs. (p. 365)

Analogy use is usually prompted by the realization that some students often lack background knowledge to learn some difficult and unfamiliar topics, such as the atom, Ohm's law, and living cells. Zeitoun (1984) has suggested the use of analogies as an interpretative bridge between the unfamiliar science topics (targets) and the knowledge that the students have.

In an analogy, an explicit comparison of the structures of two domains is done. The comparison indicates the identity of parts of the structures in the two domains—analog and target. Good analogies point out relations between attributes in the two domains. These relationships should be identical. For example, the analogous relationship between electrostatic force, F_e, between two-point charges, q_1 and q_2, a distance, r, apart, and gravitational force, F_g, between two point masses, m_1 and m_2 a distance, r, apart, is symmetrical (Gentner & Gentner, 1983). A symmetrical mapping can be made between these two expressions:

$$F_e \rightarrow F_g; k \rightarrow G; q_1 \rightarrow m_1; q_2 \rightarrow m_2; = \rightarrow =; r^2 \rightarrow r^2.$$

This is what is meant by identity of structure of the two domains.

$$F_e = k\frac{q_1 q_2}{r^2} \Leftrightarrow F_g G \frac{m_1 m_2}{r^2}, \text{ where } k \text{ and } G \text{ are constants}$$

$$\text{Analog} \qquad \text{Target}$$

It is hoped that at this point the reader has developed a full picture of what analogies are. Everyday use of the term analogy, as seen in chapter 1, includes metaphors, models, and examples. With the understanding of what an analogy is, it is important to clarify further the important distinctions among analogies, metaphors, models, and examples.

Analogies and Metaphors

Analogies and metaphors are "teaching devices" close in meaning and can easily be confused. Both analogies and metaphors express comparisons and highlight similarities. However, their difference is located in the process of comparing (Duit, 1991). An analogy compares the structures of two domains (analog and target) explicitly by pointing out the identity of parts of the structures, while a metaphor compares without explicitly pointing out the identity. In addition, metaphors often use literary representations or poetic imagery, while analogies operate at a much more analytical level. The apparent essence of a metaphor is to *hide* the grounds of comparison. A metaphor compares implicitly, highlighting features or relational qualities that do not coincide in the domains. Taken literally, metaphors are plainly false. "Viewed with the framework of [analogy] one could say that a metaphor points out some major dissimilarities in order to incite the mind to search for similarities" (Duit, 1991, p. 651; see also Ortony, 1979, pp. 174–175).

In conclusion, analogies can be seen as metaphors and metaphors as analogies (Duit, 1991). For example, a person who takes care of a given family—being responsible for the feeding, dressing, and the welfare of the entire family—is metaphorically the engine behind the family. This metaphor can be used to explain the functioning of the car engine or any machine engine. Knowing what he/she does in managing family affairs and keeping the family together, and matching with similar functions of a car engine, makes this metaphor an analogy.

Analogies and Models

Under special circumstances, models provide analogies (Coll, 2005; Duit, 1991). Ordinarily, a model is a representation in simple form or simplified description of a system or an idea. There are various types of models, including (a) *three dimensional physical representations* of objects, examination of which can ascertain facts about the objects they represent; and (b) *theoretical models*, which are descriptions of ideas (Achinstein, 1968; Asimov & Bach, 1992). Although widely used in all sciences, representational models are particularly important in engineering (Achinstein, 1968). Instead of investigating the object directly, engineers may construct a representation of it, called a prototype, which can be studied more readily and under more-controlled circumstances than the real thing.

The analog model is very important in science teaching, especially in physics. The characteristics of the prototype, which mirror those of the mother object, are not themselves reproduced. Instead, an analogy is drawn between two essentially unlike objects or systems, say X and Y, where Y represents X as its analog and Y is the model of X (Achinstein, 1968; Nashon, 2000). For Y to be the model of X, Y should be treated as something to be studied, for example, the analogy between an electric field and an incompressible fluid that flows through tubes of variable cross-sections.

In general, to speak of a model is to talk about something intended to represent a prototype, and in such a way that facts about the prototype can be ascertained by studying, making calculations with respect to, and experimenting upon the model (Achinstein, 1968; Nashon, 2000). Whereas a prototype has to be built, a model analog need not. It needs only to be described.

Theoretical models, which are sets of assumptions about X, include *the billiard ball model of a gas*, which postulates that the molecules comprising a gas exert no forces on each other except on impact (Achinstein, 1968; Nashon, 2000). Once energized, billiard balls travel in straight lines except at the instant of collision, and are very small in size compared to intermolecular distances. Other examples of models include

1. *The Bohr model of the atom*: In this model, an electron revolves round the nucleus of an atom such that its orbital angular momentum is quantized, as is the energy radiated or absorbed by the atom;
2. *The Corpuscular model of light*: In this model, light is considered to comprise particles in motion;
3. *The shell model of the atomic nucleus*: Particles in the atomic nucleus are arranged in shells. They move in orbits with quantized angular momentum in a nuclear field created by the particles in the nucleus. This model represents nuclear atomic structures. It is based on the assumption that the number of each kind of the particles (protons and neutrons) contained in each shell is given by $2(2lh + 1)$. The orbital angular momentum of particles in each shell is $lh/2$, (Where l = angular momentum and h = Planck's constant); and
4. *The free electron model of metals*: The valence electrons in the atoms of a metal move freely through the volume of the specimen. (Achinstein, 1968)

All these are examples of theoretical models. However, if a model can be used to explain unfamiliar (or new) but similar information, it becomes an analog model. For example, the Bohr atomic model can be explained using the solar planetary model. Therefore, the solar planetary model is an analog model. In other words, models may not necessarily explain the target concept but can simplify it for the purpose of explaining through an analogy. The Bohr-Rutherford atomic model does not explain how the electrons and the nucleus function; rather, it simplifies their arrangement for explaining. To explain how electrons move and how they are held in positions, another model—the solar planetary model—is used. The planetary model attributes or features include the sun, the planets, and the orbits (Brown, 1972). Matching similar attributes and identifying dissimilar attributes in both models constitutes a complete analogy.

In this analogy, the sun corresponds to the nucleus of the atom, the planets correspond to the electrons of the atom, and the orbits of the planets around the sun correspond to the paths that electrons follow (orbitals). However, there are also attributes in both models that are irrelevant. For this analogy to be complete, the irrelevant (dissimilar) attributes have to be identified. Dissimilar attributes include the moons, moons' orbits round the planets, what is inside the sun (e.g., burning hydrogen and helium gases), and the neutrons and protons in the nucleus of the atom. Although this is an oversimplified analogy, it nevertheless provides us with an example of an analogy in which an analog model (the solar planetary model) is used to explain the target model (the Bohr-Rutherford atomic model).

On the other hand, a model can be used as an analogy by itself. For example, kinetic theory, which explains the phenomena of heat and pressure as due to the motion and elastic collision of atoms and molecules, is illustrated by having a glass container with billiard balls. When heat is applied, the balls (small spheres) are seen to move randomly, colliding with each other and with the walls. Their motion increases with increase in heat supply. This arrangement is a model of what happens when heat is applied to a gas in a container, known as the "billiard ball model of gas." Although billiard balls are used, the effect is not as pronounced compared to using polystyrene spheres. This is because billiard balls are heavy. Therefore, for effective demonstration, polystrene spheres can be used because they are lighter. However, billiard balls can still be used as long as their size is smaller compared to the average intermolecular distances, because they have an advantage over glass in relation to their heat capacity: billiard balls have a lower heat capacity than polystrene spheres. At the same time, the setup is an analogy. As a model, it represents the (invisible) increase in motion of molecules and atoms and their collision with each other and the walls of the container. As an analogy, billiard balls are regarded as similar to atoms or molecules, and their randomness as similar to that of molecules, say, in a tire tube. In other words, when a phenomenon is illustrated using a model, which at the same time explains the phenomenon, then that model becomes an analogy.

A model may provide an analogy on condition that the model, which is used to explain some new concept (target), is not in the same domain as the new concept. Furthermore, the two should not be identical. For example, using the drawing of a cow's eye to explain a human eye is not an analogy; nor is drawing the model of the solar system to explain the actual solar system. But if the planetary model of the solar system is used to explain atomic structure, the relation between the two is analogous because the planetary and atomic structure models are in two different domains or paradigms (Kuhn, 1970). Therefore, the analog and target are different worlds, domains or paradigms whose functioning is perceived as similar.

This section has discussed the distinction between an analogy and a model. It has also discussed models as analogs and models as analogies. The exceptions are the few instances mentioned above when a model serves dual purposes of analogy and model. Conditions for this have been made clear through the examples discussed above. The next section discusses the relationship between analogies and examples.

ANALOGIES AND EXAMPLES

Analogies and examples serve similar purposes insofar as they are used to make the unfamiliar familiar. However, examples differ from analogies in terms of usage. An example is an instance of a concept (Duit, 1991; Glynn, 1991), with the example serving to clarify the concept by making reference to familiar objects or phenomena. Take, for example, a machine. A machine makes work easier. To make this definition clearer, an instance is necessary. Such may include a wheelbarrow, a hammer, or a tractor. All these are machines. They are examples of machines, but they are not analogies. However, examples can be used as analogs in analogies. For example, one analogous situation can be used to explain an unfamiliar analogous relationship, such as the family head and the car engine relationship being used as an example in explaining how the heart works.

The cases of metaphors as analogies, models as analogies, and examples as analogies are rare situations. Ordinarily, analogies, metaphors, models, and examples are different teaching devices used to clarify, simplify, or explain complex and new concepts or material. The aim of using such devices is to bring about the understanding of target material or concepts. The next section discusses the features of an analogy. Careful attention to these features is very important when constructing and using an analogy.

Features of an Analogy

As stated earlier, an analogy is a teaching/learning device that uses familiar material to explain unfamiliar material or to make abstract knowledge more concrete. An analogy is constituted only when the analog and target share attributes or have attributes in common. Common or similar attributes are known as *analogous attributes* (Zeitoun, 1984). Of course, part of the nature of analogy is that it may also contain dissimilar attributes or what Zeitoun calls *irrelevant attributes*. How powerful an analogy is depends on the relative number of similarities compared with the number of dissimilarities in analog and target domains (Weller, 1970).

Attributes are analogous when they exhibit

1. Structural similarity; i.e., they look similar or are arranged in a similar configuration;
2. Functional similarity; i.e., they act in a similar way and;
3. Consequential similarity; i.e., a similar or same underlying cause creates similar effects. (Zeitoun, 1984)

An analogy may have some or all of these characteristics. For example, the atom–solar system analogy has all the three characteristics. Quoting Pollard (1978), Zeitoun (1984) shows how the atom and the solar system have structural (physical) similarities because they both comprise smaller spheres orbiting larger spheres. They also have functional similarities— smaller spheres spin on axes. And both have consequential similarities since the electric force and gravity are viewed as similar causes that have effects. In addition, the electric force and gravity are similar because they both vary inversely with the square of the distance between their respective sources. However, analogies can be misunderstood and as a consequence may lead to the development of misconceptions among the learners. Stavy (1991) even warns of the resistant nature of misconceptions or preconceptions to ordinary classroom teaching. She advises that "if students are to change their ideas, they must first feel that their existing conceptions are unsatisfactory in a way" (p. 365). In most cases, students are shown only the similar attributes. As a result, students' preconceptions tend to be dominated by misconceptions (Stavy, 1991).

Teaching, to some extent, develops by building on students' correct intuitive preconceptions. Therefore, it is important that teachers look for the students' correct preconceptions rather than the incorrect. In other words, a good effective analogy has to utilize a correctly understood concept that the student already has. Personal participation of students in developing and understanding such knowledge inculcates a sense of responsibility and ownership by the learners (Polanyi, 1962). This view finds favor in Stavy's (1991) acknowledgement of the influence perceptual elements have in the support for correct intuitive knowledge, and in encouraging transfer by analogy from understanding of known cases to similar cases in the unknown. Thus, there should be at least one analogous attribute between the target (topic) and analog, otherwise the term analogy may not apply (Zeitoun, 1984).

The correct knowledge that the students already have should constitute the analog. If the analog is not well understood by the students, then it must be taught before being used to explain the target. For example, in the familiar camera-eye analogy, knowledge of the camera is used to explain the functioning of the eye. Thus, the eye constitutes the target while the camera is the analog. Knowing the main parts of the camera and how they function helps in explaining and visualizing how the eye works. Of course, some parts of the eye have to be visualized because they are not directly visible to the learner. Before the analogy is constituted, features of the analog and the target must be identified; this is then followed by matching similar attributes in the two domains. In this case, matching attributes in the camera and eye domains should be identified as shown below.

Features of the Camera

1. The focusing ring: alters the distance of the lens from the film.
2. The lens: focuses the image on the film.
3. The diaphragm: controls the amount of object light entering the camera.
4. The shutter: allows in light or shuts out light from the camera.
5. The spool: revolves as the film winds on it after every snapshot.
6. The film: receives the image of object.
7. The light-tight box: houses the camera parts (gives shape to the camera).

These attributes of the camera can be identified visually and their functioning demonstrated practically, both individually and in conjunction.

Features of the Eye

1. The sclera: acts as outer skin to maintain the shape of the eye.
2. The cornea: does the reflecting of light for the lens to focus sharply.
3. The pupil: determines the amount of light entering the eye.
4. The ciliary muscles: adjusts the thickness and curvature of the lens, which determine the focal length.
5. The iris: automatically adjusts the size of the pupil.
6. The lens: focuses the image of the object on the retina.
7. Aqueous humor: clear fluid in the eye between the lens and cornea.
8. The retina: screen of the eye on which the image is formed.
9. The optic nerve: conducts electric impulses from the retina to the brain.
10. The blind spot: point at which the optic nerve leaves the eye.
11. Vitreous humor: transparent jelly-like tissue filling the eyeball.
12. The eyelid: the upper or lower fold of skin, closing to cover the eye.

Since some of these features (attributes) of the eye are not directly visible, a part of the camera that operates in a similar manner can be used to explain them. It follows that, before this analogy is used, it is important to ascertain that the simple camera is understood well. Features that have similar functions in both the camera and the eye domains should be identified and matched first.

1. The light-tight box for the camera and the sclera for the eye.
2. The shutter for the camera and the eyelid for the eye.
3. The diaphragm for the camera and the pupil for the eye.
4. The focusing ring for the camera and the ciliary muscles for the eye.

5. The film for the camera and the retina for the eye.
6. The diaphragm adjusting ring for the camera and the iris for the eye.

Although this list does not have all the similar features in the camera and eye domains, it is probably sufficient to help students understand the working of the corresponding parts of the eye. It helps the students to visualize the "activities" in the normal eye. However, the analogy is incomplete if the dissimilar parts between the camera and the eye are ignored. The dissimilar parts are those that are unmatched; they may be extra features the analog has over the target or vice versa. Unmatched attributes in the camera include the spool; unmatched attributes in the eye include the aqueous humor, the vitreous humor, the optic nerve, and the blind spot.

In the camera–eye analogy, the unmatched parts of the eye are not explained easily. Indeed, they might each require another analogy in order to explain them. For example, the blind spot and optic nerve may be explained by comparing them to the photovoltaic cell that converts light into electric current.

Sometimes it is important to be careful not to confuse similarity with sameness. Attributes that are the same are not analogous. Zeitoun (1984) explains this by saying,

> The term analogy is not applicable when the attributes of the topic [(target)] and the analog are identical (as is the case of one human eye and the other) or literally similar (as in the case of a human eye and the eye of another mammalia, e.g., rabbit). (p. 109)

Zeitoun states that the term analogy only applies when the target and analog belong to different schemata (plural of schema). Schema is a term used by Piaget to describe the general knowledge possessed about a particular domain (Stepich & Newby, 1988; Zeitoun, 1984). For example, in the camera-eye analogy, the human eye belongs to the eye schema, whereas the camera belongs to the camera schema. Thus, the two schemata are not identical or literally similar. But they are analogically related (Zeitoun, 1984). When pictures, physical models, and other illustrations belong to the same schema as the target, they are not considered analogous (Zeitoun, 1984). The term analogy does not apply when a physical model is used to explain the working of the target in the same schema. Thus, the picture of an object is not the analog of the object when it is used as a teaching aid to explain the object because both the object and the picture belong to the same schema. Learning through analogies is therefore explained by schema theories.

Categories of Analogies

Dagher and Cossman (1992) conducted a study that explored the nature of explanations used by science teachers in junior high school classrooms. Using classroom observation of twenty public school teachers in a total of 40 class periods, Dagher and Cossman generated ten types of explanations that were conceptually related to one another. The ten categories or types of explanations generated are anthropomorphic, analogical, functional, genetic, mechanical, metaphysical, practical, rational, teleological, and tautological.

Anthropomorphism is the making of a phenomenon familiar by attributing human characteristics to nonhuman agents involved; *analogical* explanation refers to the use of a familiar situation to explain a similar unfamiliar situation; *functional* explanation refers to a situation whereby a phenomenon is explained in terms of its immediate consequence (function); *genetic* explanation refers to a situation whereby an explanation is provided by relating antecedent sequence by events, e.g., a car may fail to stop after brakes are applied because the road became slippery after an oil spill from a leaking gas tanker. Antecedent events in this case include leakage from tanker, slippery road, application of brakes, and the event being explained is failure of the vehicle to stop; *mechanical* explanation refers to a situation where causal relationships that are generally physical in nature; *metaphysical* involves making reference to supernatural agents as causes of phenomena; *practical* explanations involve instructions as to how to perform physical or mental operations; *rational* explanation involves presentation of evidence or warrant for a given claim in an effort (implicit/explicit) to compel belief; *teleological* explanation involves explaining a phenomenon in terms of how its immediate consequence (function) contributes to the probable attainment of an ultimate consequence (goal) in conjunction with other phenomena that are part of the same biological/physical system; and *tautological* explanation involves the reformulation of the how/why question or statement without adding any new information to its content (Dagher & Cossman, 1992, pp. 365–366).

From an analysis of forty transcripts recorded in classrooms of 20 seventh- and eighth-grade science teachers, Dagher (1995) identified five general categories of analogies: *compound, narrative, procedural, peripheral,* and *simple.*

Compound analogies are those involving instances where the teacher used more than one source (analog) domain to explain several ideas related to the target (target comprises several attributes explained by drawing on their similarities to several analogs). For example, to explain a complete electrical circuit, an understanding of attributes like current, resistance, potential difference, and cell or battery is essential. These are attributes of a single target—electrical circuit. The attributes could be explained using multiple analogs. For example,

- current can be explained using its similarity to water flow;
- resistance can be compared to a stone placed in the path of water flow;
- potential difference can be compared to pressure difference or difference in altitude or gradient of a landscape;
- cell or battery can be compared to a water pump.

Narrative analogies are those in which the teacher uses one source domain to explain several concepts in the target domain; for example, explaining the human body using the car as the analog. The engine from the analog (car) can be used to explain the various attributes of the human body (target). For instance, the functions of a car engine are similar to the function of the heart. When the car engine stops, the whole car stops as well. Similarly, when the heart stops functioning, the human body stops functioning too. The car engine pumps gas into its pistons using a compression shaft linked to a fuel pump. And so on.

Procedural analogies are those that pertain to procedures associated with the way in which science is carried out. These analogies do not necessarily focus on topical concepts but rather they pertain to (say) practice, conduct, and processes. For example, if a teacher were to reinforce certain laboratory procedures, he/she may invoke a certain practice in life, such as, "Do not add water to an acid but add the acid to water," especially sulphuric acid. Analogically, this is like wanting to crush a tennis ball by throwing it onto a piece of rock in front of you. To understand the comparison, the following questions may be considered:

a. What will happen to the tennis ball when it hits the piece of rock?
b. What if the piece of rock is thrown onto the tennis ball?

Similarly,

1. What happens when water is added to concentrated strong acid?
2. What happens when the concentrated acid is added to water?

The procedures in both cases have implications for the safety of the person doing the activity. In the case of a tennis ball being thrown onto a piece of rock, the ball bounces back and may hit the "thrower." And in the case of water being added to the concentrated sulphuric acid, the acid gets "spat"; that is, because of the immense heat of reaction generated, the water steams and causes "spitting" that could burn the person mixing water and acid. Because of the severe injury that is likely to result, the procedures in both cases have to be adhered to. Using one of the cases to explain or illustrate the other constitutes a procedural analogy.

Peripheral analogies are secondary or accidental analogies that depend, say, on a single and central analogy. For example, Dagher (1995) uses the case of a telephone cable having several colored transmission wires. Each wire is connected to a telephone in the home of a subscriber or telephone booth. This ensures that messages are sent to the right receivers. Similarly, the spinal cord has several nerves that link various parts of the body to the brain. Any message (action) from any part of the body must follow the correct channel (nerve) to the correct part of the brain for interpretation. Sometimes when a part of the cable system is cut, it is difficult to reconnect and therefore no messages can be exchanged between the message sender and the telephone subscriber. Similarly, when one of the nerves in the spinal cord is cut or damaged, the message exchange between the part of the body and the corresponding part of the brain is lost. With the telephone lines, the line cut is easily repaired because of the wire color coding. Without the color coding, several things can happen, including a mistake in the connection of the cut wires, which can lead to messages from a particular sender going to the wrong receiver. In a similar way, this is what can happen to a person with a faulty spinal cord. Any damaged or cut nerve cannot be retouched easily, thus leading to no reception of messages by the brain from the damaged nerve, resulting in numbness. In other words, there will be no exchange of messages between the de-linked parts of the body and the brain. The main analogy is the telephone cable-spinal cord. But a secondary analogy is introduced to explain the consequences of not coloring the telephone wire in a cable as compared to indistinctiveness of the nerves in the spinal cord. This analogy is still part of the main telephone cable-spinal cord analogy.

Simple analogies are those that require further development. Dagher (1995) gives the example of a food chain whereby a blueberry is considered to contain energy from the sun and grows until humans eat it. The plant (blueberry) stores energy from the sun just like a battery does. This analogy is considered simple not because it is easy to understand but because it requires further development. Storing energy is what characterizes the analogy; further development includes the storage process, how energy from the sun (light) is changed to starch compared to how energy from the sun is stored in the form of chemical energy.

It should be noted that, for the purpose of the explanation, irrelevant attributes (Zeitoun, 1984) or unmatched attributes (Glynn, 1991; Harrison & Treagust, 1993) have been left out. However, for an analogy to be complete, the matched and unmatched attributes should be pointed out. Leaving out either type of attributes renders the analogy incomplete.

Treagust et al. (1992) observed analogies that were either enriched or simple. By enriched analogies, Treagust et al. meant those in which teachers not only carefully and clearly showed the relationship between the analog and the target, but also dealt with the analogy's shortcomings, pointing out when misunderstandings were likely to occur. Simple analogies, as in Dagher (1995), were those that required further development.

LEARNING THROUGH ANALOGIES

Schema theories constitute paradigms that can explain certain aspects of learning through analogies. A paradigm is the thinking or framework that guides observation, interpretation, and explanation of phenomena or data (Klemke, Hobger, Rudge, & Kline, 1998; Kuhn, 1970; Mautner, 1997; Zeitoun, 1984). However, interpretation of data is central to the understanding of a paradigm (Kuhn, 1970). Thus, paradigms become what were earlier referred to as "domains" in an analogy. An analogy therefore consists of a comparison of two different paradigms or "worlds." According to Zeitoun, analogical learning is effective when *the subject transfers the analogous attributes from the analog schema to that of the target.* On how effective teaching of science should proceed, Hodson and Reid (1988) advocate the teaching of some theory before providing the students with the practical experiences. Teaching theory first provides the learners and the teacher with opportunities to theorize and explore students' already-existing ideas. In other words, practical investigations should be theory-driven, in the sense that practical investigations are used to explore, test, and develop students' ideas (or the ideas teachers are trying to persuade the students to adopt). Furthermore, according to Hodson & Reid,

> It is not that children frequently do not have the necessary and appropriate theoretical framework, but they have a different one. So they may look in the "wrong" place in the "wrong" way and make "wrong" interpretations. As a consequence they may go through the entire lesson without ever appreciating the point of the experiment, the procedure or the findings. (p. 161)

This quotation may be taken to suggest that children have ideas that may be different from the scientifically "correct" ones (accepted ideas or concepts). The aim of teaching is to help children appreciate the "correct" view of science and probably make a shift from the "wrong" ideas (misconceptions or alternative frameworks) to the "correct" ideas. The two sets of ideas are based on two different paradigms: the individual students' paradigms and the teacher's paradigm.

In an attempt to assist these children to undergo a paradigm shift, the new ideas and the methods of developing the new ideas must be plausible, intelligible, and fruitful (Hewson & Thorley, 1989; Posner et al., 1982). Once the ideas the students already hold have changed to acceptable ones, the students can use them to explain new concepts. These new ideas thus become a student's preconceptions from which analogs for analogies can be selected. However, a paradigm shift can be brought about by faulty preconceptions. Therefore, on the one hand, an analog with no errors ensures transfer of correct ideas from the analog paradigm to the target paradigm. On the other hand, a misconception in the analog will be transferred to the target leading to wrong paradigm shift. Therefore, the analog should contain "correct" science in order to explain the target "correctly."

Analogical Learning is also effective when *the subject isolates the irrelevant attributes of the analog schema from that of the target schema.* Duit (1991), quoting Kelly (1955), exemplifies this point by saying,

> Learning is…but a process of actively employing the already familiar to understand the unfamiliar. Learning, therefore, fundamentally has to do with constructing similarities between new and already known. It is precisely this aspect that emphasizes the significance of analogies in a constructivist learning approach. (p. 652)

According to Zeitoun (1984), analogical learning, which embraces constuctivist views of learning, takes place when the learners exhibit the following characteristics:

1. Familiarity with the analog,
2. No prior knowledge of the target,
3. Analogical reasoning ability,
4. Appropriate Piagetian cognitive levels (especially formal reasoning),
5. Visual imagery,
6. Cognitive complexity.

It is important that the learners be familiar and comfortable with the analog. Unfamiliar analogs may distract students' attention from studying the target and may also add a new load to the learning situation (Zeitoun, 1984). This is because the learner will be forced to understand both the analog and the target at the same time. The use of unfamiliar analogs may even drive some learners into looking for alternative, more familiar analogs, which can be a recipe for chaos or confusion in conceptual learning. According to Willner (1964), a "reason for the usefulness of the principle of analogy in constructing new scientific hypotheses is that a relationship derived from a well understood realm may be extrapolated to a dimly understood one, and provide a key for understanding" (p. 479).

Of course, the familiar will only facilitate learning to the extent that it is analogous to the topic (target). When familiar analogs are not available and easy-to-understand, unfamiliar material can be introduced to explain the target, somewhat in the style of an advance organizer (Ausubel, 1963, 1968; Ausubel, Novak, & Hanesian, 1978). Advance organizers are "typically written at the same level of abstraction, generality and inclusiveness as the learning material [targets], and achieve their effect largely through repetition, condensation [(compressing material or summarizing)], selective emphasis on central concepts and pre-familiarization of the learner with certain key words" (Ausubel, 1963, p. 214). Clearly, what matters is the link between the analog and target attributes and other conceptions that the student has. Having no prior knowledge of the target captures students' interest and desire to learn the new material. In other words, "analogies would produce best results when the learning material ... is unfamiliar, i.e., the student has not yet developed a schema for this material" (Zeitoun, 1984, p. 111).

Providing students with an analogy when they already possess relevant background knowledge about the target may interfere with their attention to learning target material. Furthermore, it may unnecessarily complicate the learning process and lead to confusion and resentment.

Analogical reasoning ability has to do with a learner's ability to extract relationships in one domain, construct a closely equivalent relationship in a different domain, and make careful inspection of both relationships to ensure that they match (Duit, 1991; Zeitoun, 1984). In other words, the learner should have the ability to understand and comprehend analogous arguments or comparisons and not to take them literally, for example, "airplane is to air as ship is to water" kind of reasoning. This way the learners become responsible for their own learning and the learning tools that facilitate this learning (Gunstone, 1994). It has, in this context, to do with the fact that learners are aware of analogies as learning devices. Not only do the learners need analogical reasoning skills, being conscious of having them assists them. The awareness and consciousness (metacognition) of having analogical reasoning skills and abilities give analogy use a prominent place among the tools of teaching science.

Learners at the Piagetian concrete operations stage (Dale, 1975) would require a bridging device to transit into the abstract. Analogies are just the right kind of devices to provide that bridge. In other words, analogies have a concretizing function. They render the unobservable (abstract) attributes perceptible by comparing them with the concrete imaginable analogs (Zeitoun, 1984). It is therefore expected that learners at the concrete operational stage can benefit from the use of analogies in studying abstract concepts or material.

Analogical learning is also enhanced by a student's ability to visualize the match between similar attributes of the analog and target. Creating the image of the match or mismatch between the analog and target attributes is a prerequisite to understanding the new (target) material. This is a high-level cognition referred to as *cognitive complexity*, one of the structural characteristics that defines individual differences. It is based on the belief that learners (people as distinct from other species, such as animals, birds, etc.) possess cognitive structures that are responsible for data processing. This structure has two components: *discriminating* structure—referring to partitioning of stimulation; and *integrating* structure—referring to how partitioned parts are related, combined, added, etc. (Zeitoun, 1984). Zeitoun states that those people classified as cognitively complex are high on the discriminative structure and/or the integrative structure. They are said to be cognitively simple if they are low on these two structural indices. The significance of cognitive analogical learning is expressed by Zeitoun as follows:

> The integrative structure might be utilized when the students relate the analogous attributes of the analog to that of the topic [(target)], the discriminative structure functions when the students isolate the irrelevant attributes between the analog and the topic [(target)]. (p. 113)

However, analogical learning, according to Zeitoun, is related to non-structural variables as well. These variables include

1. the complexity of the analogy, which refers to the number of conjoint (analogous) attributes that can be interpreted from an analogy statement,
2. the degree of concreteness of the analogy,
3. the number of analogs included in the analogy,
4. the format of presenting the analogy, which comprises two categories: mixed and separate. *Mixed* format involves a situation where both the analog and the target are presented together and compared in the lesson—matched and unmatched attributes are pointed out carefully and systematically; for example, the blind spot and the optic nerve are compared with the photovoltaic cell. *Separate* format involves a situation wherein the target and the analog are presented independently, with the comparison being made later; for example, the camera and the human eye can be presented separately with similarities and differences being discussed much later.

The teaching strategies developed above comprise three major categories: (a) student self-developed analogy strategy, (b) guided teaching strategy, and (c) expository teaching strategy in which the analogy is presented by the teacher (Zeitoun, 1984).

Whatever the approach, the aim is to have the students learn, say, correct physics concepts through the careful and correct match of similar attributes in both the analog and the target. Unfortunately, things do not always work that way. Relationships can be misinterpreted or misunderstood, leading to the development of misconceptions. In this connection, Zeitoun (1984) says that "some misconceptions might result from taking the relationship literally between the topic [(target)] and the analog. The misconceptions could be in the learning of the topic [(target)] as well as the analog" (p.118).

Several models of teaching using analogies have been suggested, among the most significant of which are the *General Model of Analogy Teaching* (GMAT) (Zeitoun, 1984) and the *Teaching With Analogies* (TWA) model (Glynn, 1991).

The General Model of Analogy Teaching (GMAT) comprises nine stages:

1. Measure some of the students' characteristics related to analogical learning in general.
2. Assess the prior knowledge of the students about the topic.
3. Analyze the learning material of the topic.
4. Judge the appropriateness of the analogy to be used.
5. Determine the characteristics of the analogy to be used.
6. Select the strategy of teaching and the medium of presenting the analogy.
7. Present the analogy to the students.
8. Evaluate the outcomes of using the analogy in teaching.
9. Revise the stages of the model.

Whereas on the surface, this model looks appealing, it has certain weaknesses. For example, the first stage does not specify which characteristics or how they can be measured. Characteristics such as analogical reasoning ability, Piagetian cognitive levels, visual imagery, and cognitive complexity (see discussion above) are not easy to measure moments before preparing a lesson. For analogical learning to be successful, teachers need a simple model that they can use without too much hassle.

Glynn (1991) claims that his Teaching With Analogies (TWA) model can be used quickly in analyzing any analogy for suitability or can be used to guide the construction of a suitable analogy. Of course, the value of an analogy is the ability to achieve its intended purpose. An analogy is good if it expresses new ideas in terms of what the students are already familiar with; it is bad if it is difficult to identify and map the important features that are similar (Glynn, 1991). It serves an explanatory purpose if among other things, (a) several features are compared, (b) most of the features compared are similar, and (c) the features compared are of conceptual significance. Glynn's TWA model is developed by utilizing these qualities (among others) and comprises six stages:

1. Introduce the target.
2. Cue retrieval of analog.
3. Identify relevant features of target and analog.

4. Map similarities.
5. Draw conclusions about target.
6. Indicate where the analogy breaks down.

This model is simple, clear, and useful. However, like GMAT, it has weaknesses, though not of the same magnitude. They are relatively easy to fix. For example, the first stage is too general. It could be elaborated by incorporating some of the ideas from GMAT, such as assessing students' prior knowledge of the target to determine whether the students already have the schemata or not. As stated earlier, analogies appear not to work well when the learners already have the schemata for the new concepts (targets). Duit (1991) reiterates the necessity of ensuring that students understand the analogy in the way the teacher thinks they should and that students see the similarities the teacher has in mind. Arguing for the case of learners having multiple meanings, Hodson (1998) says,

> If meaning is, in part, socially constructed, it follows that different social groups can negotiate and construct different meanings. Moreover, since individuals have membership of more than one social group, they need to be familiar with more than one framework of understanding and to be able to access the knowledge, language and codes of behavior appropriate to each group quickly and reliably, as the social situation changes. (pp. 74–75)

In this context of analogy use, it follows that it is important for teachers to establish common frameworks that are clear to all the students. Otherwise, if the teachers and the students operate in different frameworks, the teachers' intended meanings may not necessarily be the students' constructs.

Paulo Freire (1973), though not talking about analogies but about the role of agricultural extension workers and the peasant farmers, emphasized the need for the extension workers to enter into the farmers' cultural universe to communicate. He says, "extension agents can communicate only by entering the cultural universe of peasants. This, they can do only by becoming vulnerable and ratifying the reciprocity which their role as educators dictates" (p. xii).

Learners' frameworks constitute their cultural universe. Moreover, as a parallel to Freire's extension agents, teachers have to enter students' cultural universe of thinking and meaning making. This can happen only when the teacher appreciates the students' position. Failure to do so might lead to construction of misconceptions. Webber (1979) has voiced similar support for learner centeredness in this matter by asserting that "There is considerable evidence that the constructs, which are elicited from the subjects individually, are more personally meaningful to these subjects than are constructs applied to them from other sources" (p. 23). The reason for this is clear. Learners make sense of new information or experiences using their already-existing knowledge (Brown & Clement, 1989). When this is not the case, there is the possibility that the analog becomes confused with the theory it is serving, and when improperly used, it may serve as an obstruction to the learning process. It makes sense to point out where the analogy does not work before making conclusions about the target. Thus, Stage 6 in TWA becomes 5 and vice versa (Harrison & Treagust, 1993). By combining elements of GMAT and TWA, the researcher has proposed a simpler model, the *Working With Analogies* (WWA) model (Nashon, 2000). It comprises six steps:

1. Assess students' knowledge of analog.
2. Assess students' prior knowledge of target.
3. Identify analog and target attributes.
4. Map similar attributes.
5. Point out unmapped attributes.
6. Draw conclusions about target.

This model closely matches what Zeitoun (1984) has described as the *way of presenting an analogy*:

1. The objective of the presentation is stated first. This is done as a way of directing students' focus and attention to the topic (target).
2. The analog is introduced quickly if it is familiar to the students. It is here that the suggested model WWA is explicit—assessment of students' knowledge of the analog. Previous research has shown that analogs can often be misconceived and can lead to the propagation of misconception into the learning of new material or target. The unfamiliar analog might require some in-depth teaching until it becomes familiar to the learners.
3. A statement that initially connects the analog to the topic (target) is presented so that the students are cognitively prepared.
4. Analogous attributes are presented one at a time.
5. A transfer statement is presented.
6. Irrelevant attributes are presented.

In the WWA model, steps 5 and 6 are interchanged on the grounds that it makes sense to make a transfer statement *after* irrelevant or unmatched attributes have been identified and presented.

Teachers' Use of Analogies

Treagust et al. (1992) underscore the importance of teachers' formal education in analogy use following a study that examined how science teachers used analogies in their regular teaching routines to enable students to comprehend scientific concepts. One of the key findings was the need to have analogies founded on a well-prepared teaching repertoire of analogies using specific content in specific contexts. In addition, they concluded that teachers should view learners as constructing their own knowledge rather than being mere passive recipients of teacher-presented knowledge.

Harrison and Treagust (1993) observed a grade-10 physics teacher teaching refraction in optics using a predetermined analogy. The analogy likened a ray of light as it passed from air into glass to a pair of wheels that changed direction as they rolled obliquely from a hard onto a soft surface. The study indicated that a competent teacher could integrate this systematic approach into a teaching repertoire that can result in better conceptual understanding of phenomena among students. The study also indicated that shared attributes should be precisely identified by the teacher or students, and that the unshared attributes should be explicitly identified as well. Lastly, there was evidence of student understanding, likely to have been enhanced by systematic use of analogies (Harrison & Treagust, 1993).

Thiele and Treagust's (1994) study of four Australian chemistry teachers indicated that pictorial analogies (analogies that compare analog and target attributes using pictures) were frequently used to enhance analog familiarity and explanation, although the authors noted an absence of statements on the limitations of analogies. They also found that teachers used analogies only when in their (teachers') judgment, the students had not understood the initial explanations. In addition, there was a high frequency of analog explanation, which the authors attributed to the lack of familiarity among the students. Thus, there were no planned analogies. Rather, teachers tended to draw analogs from their own experience or professional reading and did so in response to particular (unplanned) situations.

A study by Lagoke et al. (1997), similar to the current study in many ways, aimed at determining whether the use of environmental analogies can eliminate the gender gap in science-concept attainment. It was based on the assumption that the use of analogies from the individual students' sociocultural environment can successfully act as a psychological bridge for the learning of science concepts. Students in senior secondary (eleventh grade)—205 male and 43 female—were selected from two schools in a certain township in Kaduna State, Nigeria. Using an adaptation of Glynn's TWA model, a pretest and a delayed posttest comparison showed that both male and female students attained an equivalent cognitive outcome after a 6-week treatment period; there was an indication that both boys and girls benefited significantly from teaching with environmental analogs. Environmental analogs, as used in this study, refer to analogical linkages derived from the sociocultural environment of the learner. Lagoke et al. state that

> The use of environmental analogs ensured that the introduction of new and unfamiliar concepts began from prior knowledge of the students, which bears specific relation to daily life of the students' society. Cultural traditions and beliefs in a given society were found to have exerted some effect on science teaching. (p. 376)

This study was significant for the present study because it provided a background context similar to the one in which the current study was conducted. Furthermore, except for minor differences, any study of this nature conducted in Africa is likely to reveal the anthropomorphic element of the African's worldview. Moreover, Nigeria and Kenya have gone through the same colonial experience. This colonial effect exposed the two countries to similar teaching styles, with both using English as a second language and as a medium of instruction in schools.

STUDENTS' USE OF ANALOGIES AND LEARNING

Brna (1991) shows how confrontation can be creative and useful in teaching and learning. In this study, a student first makes predictions about a particular situation; then the actual situation is presented. The intention is to incite "cognitive conflict": the student experiences difficulties in reconciling the conceptualization of the situation before and after experiencing the actual situation. This is an example of what Gunstone (1994) calls the "Predict-Explain-Observe-Explain" (PEOE) approach. It is also an example of the discrepant events strategy, in which students are presented with events that cause a conflict with the understanding they already have. Misconceptions are revealed in the predictions. The students get challenged when "correct" experiences are encountered. Brna's study indicates that the student abandons or modifies the old conceptualization of the situation in favor of the new experiences or relations (Posner et al., 1982). Brna's approach had some weaknesses. For example, putting a student in some conceptual difficulty does not necessarily make conceptual change easy or lead to abandonment of previously held ideas (Cobern, 1996; Matthews, 1998). However, the research paper provides insights into ways of identifying misconceptions that children have about certain ideas or phenomena.

Although not directly related to the use of analogies, Brna's study (1991) reinforces the idea that if the analog has misconceptions, then those misconceptions are likely to be transferred to the target in an analogy. Therefore, to resolve any apparent conceptual conflicts within the analog, it means providing the students with opportunities to express their conceptual understanding, which can then be confronted by countersituations that present plausible, intelligible, and useful solutions/explanations. This way, fewer misconceptions (or none) may be transferred to the target. It is a fact that new information learned has the potential of being used as an analog for a similar new target.

In their research to find out if pupils could use taught analogies for electric current, Black and Solomon (1987) used three groups of third-year secondary (13–14 year olds) students at the same comprehensive school. These groups were constituted from the 30th percentile downward, using the results of tests and end-of-year examinations during the students' first 2 years at the school. The three groups were taught electrical resistance, constant current flow in different parts of a circuit, and the decrease of current with increase of resistance. This was later followed by practical work with cells, ammeters, and homemade resistors of coiled wire. All the examples were confined to simple circuits, leaving the branched circuits for prediction using analogies. The first and the second groups were taught using fluid and electron analogies, respectively; the third group received instruction without use of any analogy. Prior to the teaching, everyone in the groups was asked to write about "How you imagine an electric current." This was followed by 2 hours of instruction for each group. After instruction, a written test paper, a second piece of free writing about how the pupils imagined electric current, and some selected interviews were administered. For each problem, the pupils were required to give a reason for each of their answers. In this way, the researchers were able to assess how far the pupils were using analogies in their solutions. Black and Solomon noted little difference between the groups on the familiar problems. However, the difference became significant with answers to questions that required pupils to predict what would happen in a branched circuit. Black and Solomon concluded from these findings that analogies do help pupils to predict what may happen in new situations. Once again, this study has informed the current study in many valuable ways, including elicitation and discernment of learners' ideas through testing and interviewing.

Stavy and Tirosh's (1993) study determined "when an analogy is perceived as such" by presenting seventh- to twelfth-grade students with problems related to successive division of physical and geometrical objects, and comparison of problems related to physical and geometrical objects.

From students' responses, the researchers concluded that

1. When having to solve an unfamiliar problem, students often use knowledge of a familiar problem they perceive as similar to the problem at hand;

2. The success of any given analogy depends on several factors including the way in which the teacher sets up the problem;
3. Similarity in structure of the problems determines whether students view a given problem as analogical to another one;
4. Salient external features of a problem, rather than the theoretical framework, largely influence students' responses to problems; and
5. Problems that involve the same process have a coercive effect on students' responses. This encourages students to view problems that involve the same process as identical.

In their advice to teachers and other analogy users, Stavy and Tirosh (1993) convey the importance of presenting students with structurally similar problems. They further recommend that teachers discuss with the students the validity of the analogies they make and the theoretical framework in which the analogies are embedded. Again, there is a clear indication that metacognition—students' awareness of the nature of analogs and what it is that they convey, depending on the commonly established theoretical framework—makes it easier for them to solve related problems.

Clement (1993) carried out a study that established the necessity of bridging analogies in more complex situations. This was found to be the case when dealing with content areas that were considered prone to misconceptions or alternative frameworks among students. Clement designed some lessons that used analogies to demonstrate to the students how a system of a "book-on-table" was similar to a "hand-on-spring." The purpose of this analogy was to help students appreciate the fact that the table exerts an upward force to counter the book's downward force. In other words, this analogy treats the table and the spring as matching attributes. But the results of this study show the reluctance of students to accept this premise. Clement's findings show that a bridging analogy is needed as a transition to the book-on-table/hand-on-spring analogy.

Before moving to the hand-on-spring analog, a less rigid setting was first used—flexible cardboard. This pointed to the fact that the cardboard was exerting a resistive force. Otherwise, if it did not, then the bending would not have occurred in the flexible cardboard. Thus, a table offers the same resistive force at a flat level as the flexible cardboard. The flexible cardboard system was used as a bridging analogy. Clement (1993) concluded that anchoring examples that tap intuitions that agree with accepted theories can be used as starting points because

a. Not all preconceptions are misconceptions;
b. Forming analogies between more difficult examples and an anchoring situation is an important instructional technique;

    c. Bridging by use of structural chains of intermediate analogies combined with group discussions encourages active thinking and helps students to believe in the validity of analogies;

    d. Explanatory models should be constructed from anchoring examples to provide an imageable mechanism, which explains target behavior.

Wong (1993) examined analogical reasoning in contexts where understanding is generated from loosely organized, incomplete prior knowledge rather than transferred from a well-structured domain of understanding. The results of this study support the view that analogies generated by the learner are more meaningful and enhance understanding of the target concepts better than externally generated analogies. The results also reinforce the idea that learner-generated analogies stimulate new inferences and insights. Wong concluded that individuals can productively harness the generative capacity of their own analogies in order to advance their conceptual understanding of scientific phenomena. Generative analogies are therefore considered tools to be used and modified as new understanding evolves.

In this study (Wong, 1993), because of the different backgrounds of participants (chemistry, physics, biology, geology), different analogies were generated for the same concept—the scientifically accepted account of air pressure. "Pressure is caused by collisions between colliding air molecules, and pressure inside and outside the syringe combine to create a resisting net force" (Wong, 1993, p. 1269). After experiencing this phenomenon using a gas syringe, the students were then required to generate their own analogies. Without going into the details of how the researcher went about ensuring this, it is amazing to note the kinds of analogies that the students generated, for example,

    1. The concept of "Air particles move, collide and create pressure" was explained by the following student-generated analogies:
        a. Air particles are like billiard balls in a container;
        b. Air particles are like people moving in a room;
        c. Air particles are like rubber balls inside the syringe.
    2. The concept of "Outside air pressure is an important factor" was also explained by the following student-generated analogies:
        a. Air particles are like rubber balls inside the syringe;
        b. Air is like people moving in a room;
        c. Pressure feels like a tug-of-war.

These examples illustrate the evaluative nature of analogies. To generate analogies of this kind is a sign of deep understanding of the phenomenon or concept.

Schwartz (1993) conducted two studies in which he explored whether adolescents can construct abstract visualizations to structure complex information. Using seventh-, ninth-, and tenth-grade students, Schwartz provided some students (but not others) with cues on visualization. After administering pre- and posttests, the findings revealed that providing students with instruction in visualization equipped them with the ability to develop a strategy that they could use to understand structure in complex and novel information. For example, one of the experiments required students to lay out or visualize information given in sentence form in a way that could make it easier to answer questions. The information was about the transmission of effects through a natural pathway. Schwartz says that an appropriate visualization for transmission problems integrates sentential information using three structural features:

1. It indicates the direction of transmission such that one can order a cause and effect;
2. It indicates *one-to-many* relationships such that one can infer that a single cause has multiple effects;
3. It indicates *many-to-one* relationships such that one can infer that a single effect has multiple causes.

An example of such visualization in physics relates to the energy cycle or conservation of energy. In a hydropower station, the energy stored in water in a reservoir has potential energy (Pe). This is converted to kinetic energy (Ke) when reservoir water falls onto the turbine. The turbine then turns a coil in a magnetic field to generate electric current (electrical energy-Ee). The electrical energy is transmitted through wires and may be stored in batteries as chemical energy (Ce), used in lighting (Le), heating (He), operating an electric bell (sound energy, Se), etc. An example of a visualization model for this information is shown below:

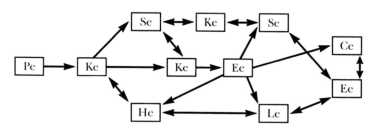

Energy Transformation Diagram

Some of the energy relationships work both ways (conversion of energy is reversible, i.e., from one form to the other and vice versa), for example,

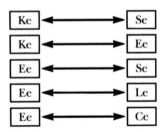

Other forms of energy can be generated from many other forms (multiple cause-single effect), for example,

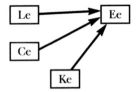

Note, Le is in the form of solar energy.

While other forms can be generated from a single cause (one cause-multiple effects), e.g.,

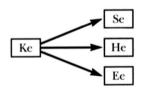

In a study titled, "Flowing Waters or Teeming Crowds: Mental Models of Electricity," Gentner and Gentner (1983) premised that particular analogies work well for certain aspects of the target, but less well for others. For example, their study showed that the water-flow analogy worked well for series electric circuits while the teeming-crowds analogy worked well for parallel electric circuits.

Usually the attributes, which constitute the structures of analogs and targets, share common relations. The bulk of analogical models used in science exhibit structure mappings between complex systems. These analogies show that relational systems hold within two different domains (analog and target). For example, if a relation, R, exists between base (analog) attributes b_i and b_j in a structural mapping M, the same relationship should exist between the target attributes t_i and t_j, i.e.,

$$M : \left[R(b_i, b_j) \right] \rightarrow \left[R(t_i, t_j) \right]$$

Systematicity rule or principle captures the intuitive view that explanatory analogies are about systems of interconnected relations. Gentner and Gentner (1983) illustrate this by using the analogy of gravitational force between point masses and the force between point charges, i.e.,

$$F_g = G\frac{M_1 M_2}{R^2} \rightarrow F_e = k\frac{Q_1 Q_2}{R^2},$$

analog   target

where G is a constant, F_g is the gravitational force between two point masses M_1 and M_2, placed a distance R apart, and F_e is the electrostatic force between two point charges Q_1 and Q_2, placed a distant R apart, with k as the constant specific to the medium in which the charges are imbedded.]

Gentner and Gentner (1983) identified two important analogies suitable for simple electricity: electricity and water analogy, and electricity and the teeming-crowd analogy. Water analogy transfers a system of relationships from hydraulics to electricity while the teeming-crowd analogy provides characteristics of a moving crowd of people in order to understand electric circuits. Two experiments were used. Experiment 1 elicited subjects' models of electric circuits by asking them whether their models predicted the types of inferences they made. In experiment 2, subjects were taught different models of electric circuits and then compared them with the subsequent patterns of inferences. The question the subjects answered was, "How does current in a simple circuit with one battery (V) and one resistor (R) compare with the current in a circuit with two resistors in series or with the batteries in parallel?"

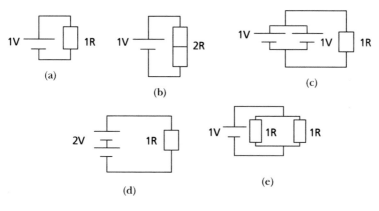

Circuit Diagrams
(Gentner & Gentner, 1983)

The results of the study showed that the subjects who drew incorrect conclusions about some of the combinations using the water analogy later drew correct conclusions using the teeming-crowd model, confirming that no single analogy can adequately explain a complex concept or has all correct properties. In this case, the water-flow analogy led to better performance on the series resistor arrangements while the teeming-crowd analogy led to better performance on the parallel resistor arrangements.

ANALOGIES AS ASSESSMENT TOOLS

Pittman (1999) provides another perspective on analogy use—that of using analogy creation as a tool of assessment. Having students generate their own analogies reveals their deeper understanding of target concepts. Pittman used student-generated analogies as an alternative assessment method to the traditional multiple choice. He administered pre- and posttests to these students with the purpose of measuring retention of factual information and gender differences. Pittman's study involved 700 seventh- and eighth-grade junior high school students. The school's ethnic composition was 90% White, 6% Asian, 1% African American, and 3% other groups. The students received instruction on "Protein Synthesis" using traditional methods. The topic was chosen because junior high students find it difficult to visualize. It should be noted that in the American context, *Asian* would refer to Chinese, Japanese, and Korean students, while in the Kenyan or British context, it would refer specifically to Indians or Pakistanis. The students were later trained by the teacher on how to think of analogies as A: B: C: D, by working through some examples of the type, "Happy is too sad as generous is to _____." The class discussed and analyzed several other teacher-generated analogies using the GMAT model. This was a way of training the students on how to generate analogies. After the training, groups of three students each generated their own analogies on subtopics of their choice under "Protein Synthesis."

The following were inferred from students' responses:

1. Student-generated analogies forced students to seek similarity relations between the information of protein synthesis and their prior knowledge.
2. The differences in subtopic selection and the type of analogy created were gender dependent.
3. Student-generated analogies were preferred by 60% of the boys who took part in the study, while 60% of the girls preferred teacher-generated analogies.

On the whole, Pittman's study (1999) showed that student-generated analogies provided a better picture of student understanding about protein synthesis than the traditional paper-and-pencil tasks, such as multiple-choice tests. This study is useful in understanding what lies beneath student-generated and teacher-applied analogies.

METACOGNITIVE THINKING: TEACHING AND LEARNING

Metacognition involves students' ability to examine what they are thinking about, to make distinctions and comparisons, to see errors in what they are thinking about and how they are thinking about it, and to make self-corrections (Ornstein & Lasley, 2000). Note that *Meta* signifies "aboutness" and is used to form new terms, which signify a discourse, the theory or field of inquiry one level above its object, which is also a discourse or field (Mautner, 1997). Metacognition is typically presented in science education literature as involving active monitoring, conscious control, and regulation of mental processes (Baird & White, 1996; Flavell, 1987; Gunstone, 1994; Mintzes, Wandersee, & Novak, 2000; Thomas & McRobbie, 2001; White, 1993; White & Frederiksen, 2005) and as such, understanding student metacognition can elucidate their learning. In this connection, Baird (1986) expresses metacognition to involve the awareness and control of one's own learning by making his/her thoughts the object of cognition. Gunstone (1994) reinforces this view by seeing metacognition to be an amalgam of students' knowledge, awareness, and control, which are relevant to their learning. Furthermore, he considers these (knowledge, awareness, and control) to be personal constructions and offers that "an appropriately metacognitive learner is one who can effectively undertake the constructive process of recognition, evaluation and where needed, reconstruction of existing ideas" (pp. 135–136). Therefore, it can be discerned from Gunstone's perspective that *knowledge, awareness, control, recognition,* and *evaluation* are in part characteristics of metacognition.

The *evaluation* construct is described by White (1992) to be "judging whether understanding is sufficient... [and] searching for connections and conflicts with what is already known" (p. 157). Thus, it is prudent to combine and interpret White's "searching for connections and conflicts with what is already known" and Gunstone's (1994) "constructive process" to mean constructing knowledge by making connections to what is familiar, hence the term *constructive connectivity* (Thomas, Anderson, & Nashon, 2008). According to Biggs (1988), metacognition involves "greater self-knowledge and task knowledge" (p. 129), which to us are characteristics of *self-efficacy* (Anderson & Nashon, 2007; Bandura, 1998; Thomas et al.,

2008). According to Bandura, *self-efficacy* refers to people's beliefs about their capabilities to produce effects.

The above synthesis demonstrates the fact that metacognition is understood differently by different people and largely seems to depend on the learning process perceived by the researcher. It is possible that each of the aspects of metacognition represented in any individual definition or clarification may be engaged differently depending on the circumstances and the learning contexts. What seems apparent is that the definitions conveyed in the above synthesis of science education literature is that the authors seem to each define metacognition on the basis of mental processes that they each perceive to illuminate the underlying processes of learning. Through this synthesis of science education literature, seven key theoretical aspects of metacognitive learning can be generated: *awareness, control, monitoring, evaluation, planning, self-efficacy,* and *constructive connectivity.* These have contributed to the development of a framework that aided our interpretation and development of deeper understanding of student metacognition and learning. However, in Thomas et al.'s (2008) SEMLI_S (Self Efficacy, Metacognition and Learning Inventory in Science), these are consolidated into five dimensions: *self-efficacy, awareness of risks to learning, control of concentration, planning-monitoring-evaluation, and constructivist connectivity* and determined in this chapter to be appropriate for signposting, interpreting, and the consequential understanding of student engagement of metacognition during a science learning discourse.

Contextualized Science Teaching and Learning

A contextual learning theory (Hull, 1993) portrays learning as occurring only when students process new information or knowledge in ways that make it meaningful in their frame of reference. According to Hull, this approach assumes that the mind naturally seeks meaning in a context by searching for relationships that make sense. Accordingly, contextual learning is organized in ways that allow students opportunities to engage in real-world problem-solving activities (Karweit, 1993). Learning in meaningful contexts has been determined to be effective (Carraher, Carraher, & Schleimer, 1985; Lave, Smith, & Butler, 1988). As Resnick (1987) has noted, decontextualizing science learning has no meaning for the students since it is seen to lack relevance outside of the school. Therefore, school field-trip venues such as science centers, museums, amusement parks, nature centers, etc. are among the important learning sites where students can be allowed and encouraged to experience relevance (Griffin & Symington, 1997). Here, learning means the processes of recognition, evaluation, and revision of personal conceptual frameworks by the learner, which develops continuously within and across a multiplicity of socially situated settings (Bruner, 1996; Driver, 1983; Gergen, 1995; Lave, 1988; Mintzes & Wandersee, 1998; Mintzes, Wandersee, & Novak, 1997). There is a large body of literature that demonstrates the value of learning in out-of-school settings (e.g., Anderson, Lucas, & Ginns, 2003, Anderson, Lucas, Ginns, & Dierking, 2000; Ramsey-Gassert, Walberg, & Walbert, 1994; Rennie & McClafferty, 1996). Researchers, including Gerber, Cavallo, and Merek (2001) and Ellenbogen (2003) have investigated learning in such settings in relation to students' cognitive processes, including their scientific reasoning abilities, and consequent conceptual learning by indicating great promise in the ways in which learning is reinforced across multiple contexts.

REFERENCES

Achinstein, P. (1968). *Concepts of science: A philosophical analysis.* Baltimore: The Johns Hopkins Press.

Aikenhead, G. S. (1996). Science education: Border crossing into the subculture of science. *Studies in Science Education, 27,* 1–52.

Aikenhead, G. S., & Ogawa, M. (2007). Indigenous knowledge and science revisited. *Cultural Study of Science Education, 2,* 539–620.

Anderson, D., Lucas, K., & Ginns, I. (2003). Theoretical perspectives on learning in an informal setting. *Journal of Research in Science Teaching, 40*(2), 177–199.

Anderson, D., Lucas, K. B., Ginns, I. S., & Dierking, L. D. (2000). Development of knowledge about electricity and magnetism during a visit to a science museum and related post-visit activities. *Science Education, 84,* 658–679.

Anderson, D., & Nashon, S. (2007). Predators of knowledge construction: Interpreting students' metacognition in an amusement park physics program. *Science Education, 91*(2), 298–320.

Asimov, I., & Bach. D. F. (1992). *Atom–Journey across the subatomic cosmos.* Penguin USA.

Ausubel, D. P. (1963). *The psychology of meaningful verbal learning.* New York: Grune & Stratton.

Ausubel, D. P. (1968). *Educational psychology: A cognitive view.* New York: Holt, Rinehart, & Winston.

Ausubel, D., Novak, J., & Hanesian, H. (1978). *Educational psychology: A COGNITIVE View* (2nd ed.). New York: Holt, Rinehart & Winston.

Baird, J. R. (1986). Improving learning through enhanced metacognition: A classroom study. *European Journal of Science Education, 8*(3), 263–282.

Baird, J. R., & White, R. T. (1996). *Metacognitive strategies in the classroom.* London: Teachers College Press.

Bandura, A. (1998). Self-efficacy. In V. S. Ramachandran (Ed.) *Encyclopedia of human behavior* (Vol. 4, pp. 71–81). New York, NY: Academic Press.

Biggs, J. B. (1988). The role of metacognition in enhancing learning. *Australian Journal of Education, 32*(2), 127–138.

Black, D., & Solomon, J. (1987). Can pupils use taught analogies for electric current? *School Science Review, 69*(247), 249–254.

Bloom, J. W. (1995). Assessing and extending the scope of children's context of meaning: Context map as methodological perspective. *International Journal of Science Education, 17*(2), 167–187.

Brna, P. (1991). Promoting creative confrontations. *Journal of Computer Assisted Learning, 7*(2), 114–122.

Brown, D. E., & Clement, J. (1989). Overcoming misconceptions via analogical reasoning: Abstract transfer versus explanatory model construction. *Instructional Science, 18,* 237–261.

Brown, P. L. (1972). *Astronomy in colour.* London: Blandfbrd Press.

Bruner, J. (1996). *The process of education.* Cambridge, MA: Harvard University Press.

Carey, S. (2000). Science education as conceptual change. *Journal of Applied Development Psychology, 21*(1), 13–19.

Carraher, T., Carraher, D., & Schleimer, A. (1985). Mathematics in the streets and in schools. *British Journal of Developmental Psychology, 3,* 21–29.

Claxton, G. (1993). Minitheories: A preliminary model of learning science. In P. Black & A. Lucas (Eds.), *Children's informal ideas in science* (pp. 45–61). New York: Routledge.

Clement, J. (1993). Using bridging analogies and anchoring intuitions to deal with students' preconceptions. *Journal of Research in Science Teaching, 30,* 1241–1257.

Cobern, W. W. (1996). Worldview theory and conceptual change. *Science Education, 80*(5), 574–610.

Coll, R. (2005). The role of models/and analogies in science education: Implications from research. *International Journal of Science Education, 27*(2), 183–198.

Dagher, Z. R. (1995). Analysis of analogies used by science teachers. *Journal of Research in Science Teaching, 32*(3), 259–270.

Dagher, Z., & Cossman, G. (1992). Verbal explanation given by science teachers: Their nature and implications. *Journal of Research in Science Teaching, 29*(4), 361–374.

Dale, L. G. (1975). Some implications from the work of Jean Piaget. In P. L. Gardner (Ed.), *The structure of science education,* 115–139. Hawthorn, Victoria, BC: Longman.

Driver, R. (1983). *The pupil as scientist.* Milton Keynes, UK: Open University Press.

Driver, R. (1989). Students' conceptions and the learning of science. *International Journal of Science Education, 11,* (5), 481–490.

Driver, R., & Erickson, G. (1983). Theories in action: Some theoretical and empirical issues in the study of students' conceptual frameworks in science. *Studies in Science Education, 10,* 37–60.

Duit, R. (1991). On the role of analogies and metaphors in learning science. *Science Education, 75*(6), 649–672.

Duit, R., & Treagust, D. (2003). Conceptual change: A powerful framework for improving science teaching and learning. *International Journal of Science Education, 25*(6), 671–688.

Ellenbogen, K.M. (2003). *From dioramas to the dinner table: An ethnographic case study of the role of science museums in family life.* Dissertation Abstracts International, 64(03), 846A. (University Microfilms No. AAT30-85758).

Flavell, J. (1987). Speculations about the nature and development of metacognition. In F. E. Weinert & R. H. Kluwe (Eds.), *Metacognition, motivation, and understanding* (pp. 21–29). Hillsdale, NJ: Lawrence Erlbaum.

Freire, P. (1973). *Education for critical consciousness*, New York: The Continuum Publishing Company.

Gentner, D., & Gentner, D. R. (1983). Flowing waters or teeming crowds: Mental models of electricity. In D. Gentner & A. L. Stevens (Eds.), *Mental models* (pp. 99–129). Hillsdale, NJ: Lawrence Erlbaum Associates.

Gerber, B. L., Cavallo, A. M. L., & Marek, E. A. (2001). Relationships among informal learning environments, teaching procedures and scientific reasoning ability. *International Journal of Science Education, 23*(5), 535–549.

Gergen, K. J. (1995). Social construction and the educational process. In L. Steffe & J. Gale (Eds.). *Constructivism in education* (pp. 17–39). New Jersey: Lawrence Erlbaum Associates, Inc.

Glynn, M. S. (1991). Explaining science concepts: A teaching with analogy model. In M. S. Glynn, H. R. Yeany, & K. B. Britton (Eds.), *The psychology of learning science* (pp. 219–240). Hillsdale, NJ: Laurence Erlbaum.

Griffin, J.,& Symington, D. (1997).Moving from task-oriented to learning-oriented strategies on school excursions to museums. *Science Education, 81*(6), 763–779.

Gunstone, R. F. (1994). The importance of specific science content in the enhancement of metacognition. In P. J. Fensham, R. F. Gunstone, & R. T. White (Eds.), *The content of science: A constructivist approach to teaching and learning* (pp. 131–146). Washington DC: Falmer Press.

Harrison, A. B., & Treagust, D. F. (1993). Teaching with analogies: A case study in grade 10 optics. *Journal of Research in Science Teaching, 30*(10), 1291–1301.

Harrison, A. G., & Treagust, D. F. (2000). Learning about atoms, molecules, and chemical bonds: A case study of multiple-model use in grade 11 chemistry. *Science Education, 84,* 352–381.

Hewson, P. W., & Thorley, N. R. (1989). The conditions for conceptual change in the classroom. *International Journal of Science Education, 11,* 541–553.

Hodson, D. (1998). *Teaching and learning science: Towards a personalized approach.* Philadelphia: Open University Press.

Hodson, D., & Hodson, J. (1998a). From constructivism to social constructivism: A Vygotskian perspective on teaching and learning science. *School Science Review, 79,* 33–40.

Hodson, D., & Hodson, J. (1998b). Science education as enculturation: Some implications for practice. *School Science Review, 80*(290), 17–24.

Hodson, D., & Reid, D. (1988). Changing priorities in science education, part II. *School Science Review, 70*(251), 159–165.

Htano, G., & Inagaki, K. (2003). When is conceptual change intended? A cognitive-sociocultural view. In G. M. Sinatra & P. R. Pintrich (Eds.), *Intentional conceptual change* (pp. 407–427). Mahwah, NJ: Lawrence Erlbaum Associates Publishers.

Hull, D. (1993). *Opening minds, opening doors: The rebirth of American education.* Waco, TX: Center for Occupational Research and Development.

Karweit, D. (1993). *Contextual learning: A review and synthesis.* Baltimore, MD: Center for the Social Organization of Schools, Johns Hopkins University.

Kelly, G. A. (1955). *The psychology of personal constructs* (Vols. 1–2). New York: Norton.

Klemke, E. D., Hollinger, R., Rudge, D. W., & Kline, D. (*1998*). *Introductory readings in the philosophy of science.* New York: Prometheus Books.

Kilbourn, B. (2000). *For the love of teaching.* Mahwah, NJ: Lawrence Erlbaum.

Kuhn, T. S. (1970). *The structure of scientific revolutions.* Chicago: Chicago University Press.

Lagoke, B. A., Jegede, O. J., & Oyebanji, P. K. (1997). Towards an elimination of the gender gulf in science concept attainment through the use of environmental analogs. *International Journal of Science Education, 19*(4), 365–380.

Lave, J. (1988). *Cognition in practice.* Cambridge: Cambridge University Press.

Lave, J., Smith, S., & Butler, M. (1988). Problem solving as an everyday practice. In *Learning mathematical problem solving* (Report IRL 88-0006). Palo Alto, CA: Institute for Research and Learning.

Limon, M. (2001). On the cognitive conflict as an instructional strategy for conceptual change: A critical appraisal. *Learning and Instruction, 11,* 357–380.

Matthews, M. R. (1994). *Science teaching: The role of history and philosophy of science.* London: Routledge.

Matthews, M. R. (Ed.) (1998). *Constructivism in science education.* Dordrecht: Kluwer Academic Publishers.

Mautner, J. (Ed.). (1997). *Dictionary of philosophy.* London: Penguin Books.

Mintzes, J. J., & Wandersee, J. H. (1998). Reform and innovation in science teaching: A human constructivist view. In J. J. Mintzes, J. H. Wandersee, & J. D. Novak (Eds.), *Teaching science for understanding: A human constructivist view* (pp. 30–58). San Diego, CA: Academic Press.

Mintzes, J. J., Wandersee, J. H., & Novak, J. D. (1997). Meaningful learning in science. The human constructivist perspective. In G. D. Phye (Ed.), *Handbook of academic learning: Construction of knowledge* (pp. 405–447). San Diego, CA: Academic Press.

Mintzes J. J., Wandersee, J. H., & Novak, J. D. (Eds.). (2000). Assessing science understanding: A human constructivist view. *Educational Psychology Series.* San Diego, CA: Academic Press, Inc.

Nashon, S. M. (2000). Teaching physics through analogies. *OISE Papers in STSE Education, 1,* 209–223.

Nashon, S. M. (2004). The nature of analogical explanations high school physics teachers use in Kenya. *Research in Science Education, 34,* 475–502.

Nashon, S. M. (2006). A proposed model for planning and implementing high school physics instruction. *Journal of Physics Teacher Education Online. 4*(1), 6–9.

Nashon, S., & Anderson, D. (2004). Obsession with "g": A metacognitive reflection of a laboratory episode. *Alberta Journal of Science Education, 36*(2), 39–44.

Ornstein, A. C., & Lasley, T. J. (2000). *Strategies for effective teaching.* Toronto: McGraw Hill.

Ortony, A. (1979). Beyond literal similarity. *Psychological Review, 86*(3), 161–180.

Piaget, J. (1970). *Science of education and the psychology of the child.* New York: Viking.

Pittman, K. M. (1999). Student-generated analogies: Another way of knowing? *Journal of Research in Science Teaching, 30*(1), 1–22.

Polanyi, M. (1962). *Personal knowledge.* Chicago: The University of Chicago Press.

Polanyi, M. (1974). *Personal knowledge: Towards a post-critical philosophy.* Chicago: Chicago University Press.

Pollard, J. M. (1978). Monte Carlo methods for index computation (mod p). *Mathematics of Computation, 32,* 918–924.

Posner, G. J., Strike, K. A., Hewson, P. W., & Gertzog, W. A. (1982). Accommodation of scientific conception: Toward a theory of conceptual change. *Science Education, 66*(2), 221–227.

Ramey-Gassert, L., Walberg III, H. J., & Walberg, H. J. (1994). Re-examining connections: Museums as science learning environments. *Science Education, 78*(4), 345–363.

Rennie, L. J., & McClafferty, T. P. (1996). Science centres and science learning. *Studies in Science Education, 27,* 53–98.

Schwartz, D. L. (1993). The construction and analogical transfer of symbolic visualizations. *Journal of Research in Science Teaching, 30*(10), 1309–1325.

Stavy, R. (1991). Using analogy to overcome misconceptions about matter. *Journal of Research in Science Teaching, 28*(4), 305–313.

Stavy, R., & Tirosh, D. (1993). When analogy is perceived as such. *International Journal of Science Teaching, 30*(10), 1229–1239.

Stepich, D. A., & Newby, T. J. (1988). Analogizing as an instructional strategy. *Performance and Instruction, 27*(9), 21–23.

Thiele, R. B., & Treagust, D. F. (1994). An interpretive examination of high school chemistry teachers' analogical explanations. *Journal of Research in Science Teaching, 31,* 227–242.

Thomas, G., Anderson, D., & Nashon, S. (2008). Development and validity of an Instrument designed to investigate elements of science students' metacognition, self-efficacy and learning processes: SEMLI-S. *International Journal of Science Education, 30*(13), 1701–1724

Thomas, G. P., & McRobbie, C. J. (2001). Using a metaphor for learning to improve students' metacognition in the chemistry classroom. *Journal of Research in Science Teaching, 38,* 222–259.

Treagust, D. F., Duit, R., Joslin, P., & Landauer, I. (1992). Science teachers' use of analogs: Observations from classroom practice. *International Journal of Science Education, 14*(4), 413–422.

Webb, J. M. (1985). Analogies and their limitations. *School Science and mathematics, 85*(8), 645–650.

Webber, J. (1979). *Personal construct theory: Concepts and applications.* New York, NY: Wiley.

Weller, A. (1970). The role of analogy in teaching science. *Journal of Research in Science Teaching, 7,* 113–119.

White, B. (1993). Thinker tools: Causal models, conceptual change, and science education. *Cognition and Instruction, 10*(1), 1–100.

White, B., & Frederiksen, J. (2005). Cognitive models and instructional environments that foster young learners' metacognitive development. *Educational Psychologist, 40,* 211–223.

White, R. T. (1992). Implications of recent research on learning for curriculum and assessment. *Journal of Curriculum Studies, 24*(2), 153–164.

White, R. T., & Gunstone, R. F. (1992). *Probing understanding.* London: Falmer Press.

Willner, A. (1964). Experimental analysis of analogical reasoning. *Psychological Reports, 15,* 479–494.

Wong, E. D. (1993). Understanding the generative capacity of analogies as a root for explanation. *Journal of Research in Science Teaching, 30*(10), 1259–1272.

Zeitoun, H. H. (1984). Teaching scientific analogies: A proposed model. *Research in Science and Technological Education, 2,* 107–125.

Dr. Samson Madera Nashon is Associate Professor of Science Education at the University of British Columbia, Canada. He generally researches ways of teaching and learning science. His area of research specialization focuses on students' alternative understandings that have roots in cultural backgrounds and curricula, and are accommodative of students with varying degrees of abilities. Dr. Nashon's research is dominantly guided by contemporary theories of constructivism. His ongoing and most recent research projects include investigating Teacher Pedagogy and School Culture: The Effect of Student Learning on Science Teachers' Teaching and Culture; Students' Ways of Knowing; and Metacognition and Reflective Inquiry. Besides engaging in research, Dr. Nashon teaches educational research methods including Action Research in addition teaching both graduate and undergraduate science education courses. http://www.edcp.educ.ubc.ca/content/samson-nashon.

J. Douglas Adler works in the area of Elementary Science Education at the University of British Columbia. He teaches in the Teacher Education program using an Inquiry approach. His area of research specialization focuses on teacher candidates and their pedagogical content knowledge in teaching elementary science.

CHAPTER 6

CONCEPT MAPPING AND THE TEACHING OF SCIENCE

Sadiah Baharom
Sultan Idris University of Education
Malaysia

ABSTRACT

This chapter will explain the various aspects of concept mapping and its application in the teaching of science. The introduction to concept mapping will encompass its definition, characteristics, and structure. An underlying theory and the psychological foundations that supports the use of concept mapping to attain meaningful learning will be discussed. Two important theories that will be elaborated is Ausubel's assimilation learning theory for meaningful learning and Novak's human constructivist view on concept mapping. An important aspect of concept mapping is its application in teaching and learning. This aspect will discuss the advantages of using concept mapping in a variety of situations in teaching and learning. A detailed account on the construction of concept maps, their various formats and types will be given to assist teachers and students to teach and learn science. A crucial aspect that has always been neglected is the assessment of concept maps. This chapter will clarify how concept maps can be quantitatively and qualitatively assessed. This chapter closes with several issues pertaining to the use and application of concept maps in teaching and learning science.

Contemporary Science Teaching Approaches, pages 115–135

INTRODUCTION

Most students rely on learning strategies that have worked well for them in achieving good grades in high-stake exams. These strategies often include rote learning, passive learning, memorization, and recall of facts. This tendency is due to the fact that school assessment captures little more than the student's acquisition of facts, problem-solving algorithms, and concept definitions. Little attention is being paid to whether or how students can build meaningful knowledge structures in an effort to achieve quality learning.

Gaining meaningful learning entails a deliberate effort from educators to implement the best strategy in students' concept acquisition. An efficient instructional strategy that can support and encourage meaningful learning is through concept mapping. Novak (2003) strongly proposes the use of concept maps in promoting meaningful learning and assessing learning outcomes.

According to Novak (2003), rote-mode learners get trapped into a cycle of memorizing whatever they can and hoping this will be sufficient to successfully pass the examinations. However, this results in them failing to build the knowledge structure that could permit them to achieve the quality of learning intended, which is to accomplish meaningful learning and gain success and confidence in future learning and novel problem solving.

What is a Concept Map?

Novak and Gowin (1984) coined the term concept map, which they view as a tool to "tap into a learner's cognitive structure and to externalize, for both learner and teacher to see, what they already know" (p. 40). Based on Ausuble's (1963) theory, which put forward a hierarchical cognitive structure and the principles of progressive differentiation and integrative reconciliation, Novak and Gowin argued that concept maps should be (a) hierarchical in nature with superordinate concepts at the top; (b) labeled with appropriate linking words; and (c) cross-linked such that relations between subsections of the hierarchy can be identified. The hierarchical structure arises because "new information often is related to and subsumable under more general, more inclusive concepts" (p. 97). The hierarchy expands according to the principle of progressive differentiation where new concepts and new links are added to the hierarchy, either by creating new branches or by differentiating existing ones further. These cross-links—between one segment of the concept hierarchy and another segment—represent the integrative connection among subdomains of the structure (Novak & Gowin, 1984).

A concept map is thus a structural representation of concepts in a certain subject domain. This structural representation consists of nodes and labeled lines, where nodes represent concepts and the labeled lines tell how two concepts are related. When two nodes are connected by a labeled line, it forms a proposition. According to Dochy (1994), a proposition is the basic unit of meaning in a concept map and the smallest unit that can be used to judge the validity of the relation (labeled line) drawn between two concepts. The concept map structure thus claims to represent the aspects of students' declarative knowledge in a certain subject domain.

According to Novak and Gowin (1984), a concept map is a schematic device representing a set of concept meanings, which is embedded in a framework of proposition. These maps are drawn as node-link-node representations, where the nodes depict important concepts and links denote the relation between the pair of concepts. Added to this, the labels on the links represent the nature of the relationship. In a concept map, any two concepts that are linked by a labeled line is known as a proposition. Propositions are in essence the meaningful statements of concepts. A graphical representation of this explanation is shown in Figure 6.1.

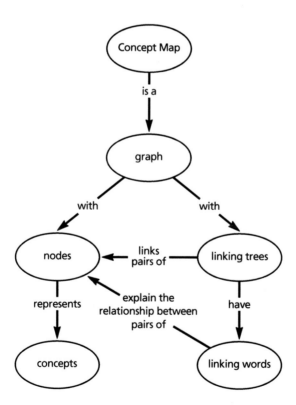

Figure 6.1   A concept map of a concept map.

Thus, concept maps can be construed as providing a picture of how students mentally organize, structure, and represent key concepts in a certain domain of knowledge. It is also a metacognitive tool that gives an insight into the minds of the learners, depicting their knowledge structure and understanding. It is a knowledge map that is a graphical or structural representation of a student's declarative content knowledge (Osmundson, et al., 1999; Ruiz-Primo, Shavelson, & Shultz, 1997). Logically, well-connected, organized concept maps should characterize experts, while an isolated, disassociated map typifies novice learners (Shavelson & Ruiz-Primo, 1998).

Theoretical Underpinnings of Concept Mapping

The construction of concept maps mediates meaningful learning. Ausuble's (1963) Meaningful Reception Learning Theory is concerned with how students learn large amounts of meaningful materials from verbal or textual presentations. He places strong emphasis on what students already know as the basis for new learning. His view is clearly stated: "If I had to reduce all of educational psychology to just one princple, I would say this: The most important single factor influencing learning is what the learner already knows. Ascertain this and teach him accordingly."

This principle has now become fundamental to understanding how people learn and construct new knowledge. According to Ausuble, in order for students to learn meaningfully, they must integrate new knowledge with what they already know. The new concepts learned must be integrated or incorporated into more inclusive concepts or ideas.

According to Novak (1990), meaningful learning happens when potentially meaningful propositions are subsumed under more inclusive ideas in existing cognitive structures. This process is called subsumption. The new propositional meanings are hierarchically organized with respect to the level of abstraction, generability, and inclusiveness. The concept mapping technique improves meaningful learning by allowing students to graphically represent and arrange concepts in a hierarchy. This hierarchical arrangement would also aid learners to progressively differentiate among concepts. This is the process of progressive differentiation in which learners differentiate between concepts as they learn more about them. During the process of integrative reconciliation, the learner reorganizes relationships between concepts and does not compartmentalize them (Novak, 1990).

The concept map thus becomes a tool that science teachers can use to determine the nature of students' existing ideas. The map can be used to make explicit the key concepts to be learned and suggest linkages between the new information to be learned and what the students already know. Teachers can use a concept map preceding instruction as a tool to generate meaningful discussion of students' ideas. Following initial construction and discussion, instructional activities can be designed to explore alternative frameworks of knowledge, resulting in cognitive accommodation.

The most fundamental basis in Ausubel's assimilation theory is the distinction between rote and meaningful learning. Rote memory is used to remember sequences of objects but does not aid the learner in understanding relationships between objects. Rote learning happens when learners make little or no effort to relate new information received to the already existing relevant knowledge she or he already possesses.

On the contrary, meaningful learning takes place when the learner consciously relates and incorporates new information into the relevant knowledge structures she or he already possesses. Achieving meaningful learning is an important step in ensuring that the quality of knowledge acquired will benefit students in the application of concepts and subsequent learning.

Uses of Concept Mapping

Concept maps are useful tools in portraying students' learning. Some of the uses of concept maps include:

Encouraging Deep and Meaningful Learning

Concept mapping encourages deep learning where learners consciously seek underlying meanings and connections and are personally involved in the learning task. This technique allows students to see connections between ideas they already possess, make new connections with newly learned concepts, and organize information in a logical manner that allows future concepts to be included. The concept map is also a useful and practical tool to assess the quality of students' science-concept acquisition. Through concept maps, depth of understanding, validity of knowledge, complexity of knowledge structures developed, and misconceptions can be easily identified.

Research findings have revealed that concept mapping can significantly improve the quality of learning. Concept mapping strategies have also been proven powerful for eliciting, capturing, and achieving knowledge. According to Edmonson (2000), students who learn meaningfully relate information from different sources in an attempt to integrate what they have learned with the intention of imposing meaning. Patterns and links within a concept map pinpoint its owner's understanding and indicates the student's readiness to progress in a certain direction (Kinchin, Hay, & Adam, 2000). Concept maps act as a tool to focus discussion. The benefit of focusing on a map during discussion is that it reduces strain on the working memory of learners (Kinchin, 2000).

Tool for Representing Knowledge

The concept map is an ideal tool for representing the knowledge that students have acquired over time. This is based on Ausubel's belief that students' preexisting cognitive structures are the anchor for all new information, and that concepts derive their meanings from links with other concepts in this preexisting structure (Arnaudin & Mintzes, 1985). Concept maps also have the ability to represent different aspects of students' understanding (Shavelson et. al., 1993). The nodes represent the initial concepts that are already present in students' minds. The links that students' make between nodes represents the contextual knowledge that the students have (Shavelson et. al., 1993).

According to Kinchin (2000), the gross morphology of the maps is also indicative of the students' learning processes and progress. In his qualitative evaluation of concept maps, Kinchin classified these maps as having specific structures, which he named spokes, chains, and nets (Figures 6.2 and 6.3).

	SPOKE	CHAIN	NET
Structure			
Hierarchy	Single level	Many levels, but often inappropriate	Several justifiable levels
Additions	Additions to central concept do not interfere with associated concepts	Cannot cope with additions near the beginning of the sequence	Additions/deletions may have varying influence as "other routes" are often available through the map
Deletions	Have no effect on overall structure	Disrupt the sequence below the deletion	
Links	Often simple	Often "compound," only making sense when viewed in the context of the previous link	Often employ technical terminology to enhance meaning

Figure 6.2 Characteristics of spoke, chain, and net-type concept maps (Kinchin & Hays, 2005).

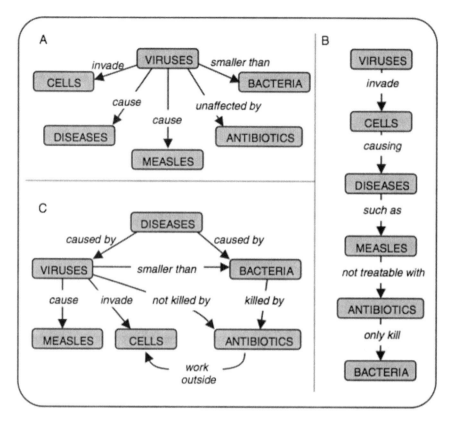

Figure 6.3 Exemplar concept map fragments of six concepts associated with pathogenic microbes. A, spoke; B, chain; C, net (Kinchin & Hay, 2005).

Kinchin (2000) also identified other indicators of "expertise" such as connectedness, link quality, and link variety. Table 6.1 outlines the differences between novice and expert maps.

Jonassen (1997) proposed that the more concepts to which a given concept is linked, the better defined that concept is. This signifies that, the more dense the network of links, the better is the quality of thinking that takes place. This argument is supported by research that showed that novices and experts structure their knowledge in different ways (Chi et al., 1988). Novak and Gowin (1984) further emphasize that experts tend to display "conceptually rich tapestries of interrelated ideas," while novices tend to possess undifferentiated, incomplete, and sometimes erroneous knowledge structures. Furthermore, experts appear to make efficient use of their dense networks, while novices tend to portray their thinking in disorganized arrays (Stevens et al., 1996).

TABLE 6.1 Differences in Features of the Concept Maps Drawn by Novice and Expert Students

Novice	Experts
Connectedness	
Disjointed structure dominated by linear arrangement in isolated clusters	Highly integrated structure with numerous cross-links
Link quality	
Links are often inappropriate. Usually single words that add little to meaning and using nonspecialist terminology	Appropriate linking phrases that add to meaning of concepts, using specialist language of the domain
Link variety	
The same linking words are used for a number of links, suggestive of narrow-range thought processes	Diversity of linking phrases illustrating a range of thought processes
Dynamism	
Stable over time, suggesting a lack of active engagement in knowledge restructuring	Changes over time, reflecting active interactions with alternative knowledge structures
Concepts	
Concentration on specific concepts, indicating a limited perspective	Concentration on major overarching concepts to create an overview

Source: Kinchin (2000).

Tool for Note Taking

> Concept Maps abandon the list structure of conventional note-taking completely in favor of a two dimensional structure. A good Concept Map shows the "shape" of the subject, the relative importance of information and ideas, and the way that information relates to other information. Typically Concept Maps are more compact than conventional notes, often taking up one side of paper. This helps associations to be made easily. (Jonassen, Carr, & Yeuh, 1998)

Concept maps can represent information in a comprehensive and clear manner. This allows for information to be gathered and communicated easily. The concept map also structures the information so that it could be easily accessed, quickly understood, and shared with others.

As a Collaborative Tool

According to Okada and Simon (1997), students are more likely to entertain various hypotheses and explore different ideas when working in groups. Concept mapping is a technique that can involve many learners, especially in the preliminary stage of brainstorming keywords for a particular topic. All learners involved in the group will contribute in listing keywords and organizing them on the map. Collaboration is achieved when each individual is able to criticize and modify the map while learning from others in the same group.

As a Creativity Tool

Drawing a concept map can be compared to participating in a brainstorming session. Ideas put down on paper become clearer, and the mind becomes free to accept new ideas. New ideas that are linked to previous ideas may trigger new associations leading to other ideas.

As a Learning Tool

Constructivist learning theory emphasizes the notion that new knowledge should be integrated into existing structures in order to be remembered and acquire meaning. Concept mapping encourages this process by making it explicit and making the learner consciously pay attention to the relationship between concepts. Jonassen (1996) argues that students display their best thinking ability when trying to represent something graphically, and thinking is a necessary condition of learning.

Concept mapping is also favorable as a tool for problem solving where it is used to enhance the problem-solving phases of generating alternative solutions or options. When working to solve problems in small groups, learners also benefit from the communication-enhancing properties of concept mapping.

Advantages of Concept Mapping

Some of the advantages of concept maps include

1. Concept mapping promotes active learning since learners must construct their own individual map when learning. In this manner, learners consciously think about their own organization of knowledge and make decisions as how to explicitly display it on paper.
2. A concept map organizes information by grouping related concepts in a structure that depicts the hierarchical nature of a particular domain and illustrates relationships between concepts through cross-linking.
3. Being graphical in nature, concept maps offers easy visualization of knowledge, thereby fostering long-term retention of concepts learned. This facilitates faster and more effective review of topics learned.
4. Branching and cross-linking in concept maps provide a nonlinear focus toward learning. This aspect nurtures learners' ability for think-

ing on a higher level, such as making comparisons between groups, looking for interrelationships, and developing a holistic understanding of the content area.

5. Concept mapping helps learners develop the ability to determine their own learning needs, and this encourages self-directed learning abilities and contributes to learners lifelong learning skills.

Disadvantages of Concept Mapping

1. Learners need time to learn the correct concept mapping technique if meaningful learning is to be achieved. This takes time and skill on the part of the learners.
2. It also takes a good portion of learning time if learners need to map out concepts learned. Learners will take shorter time reading instead of mapping.
3. Concept mapping can be laborious and make students tired.

CONSTRUCTING A CONCEPT MAP

Basically, there are six steps in constructing a concept map:

Step 1: Identifying Concepts

Identify and list all terms and concepts associated in the topic of study. Learners can form groups to brainstorm on terms that are related to the understanding of the concepts learned. These terms can be written on Post-it notes to ease arrangement on the map. Generate as many related terms as possible.

Step 2: Organizing Concepts

Arrange concepts by sticking them on a whiteboard or a large piece of paper. Create groups or subgroups of concepts that are related to each other. Look out for hierarchies. Place more-general concepts at the upper portion followed by more-specific concepts lower down the hierarchy. Do not use the same concept in more than one group; instead use a linking line to make a connection.

Step 3: Positioning Concepts

Arrange concepts to show your collective understanding of how each concept is related to each another. Place closely related concepts together. Work out connections within a group before looking for intergroup connections. Place the most general and important concept at the top of the hierarchy and arrange less-general concepts below it. Rearrange concepts as you go along, bearing in mind how each concept links to the other. Remember that your concept map need not be symmetrical. One portion of the map could have more concepts than others. Each learners' concept map is unique and different from other learners.

Stage 4: Linking Concepts

Draw lines with arrows connecting one concept to another. Write a word or short phrase to explain the connection between these concepts. This word should specify the relationship between the connected concepts. More than one arrow can originate or end on a particular concept.

Stage 5: Revising Concepts

Examine the concept map drawn. Discuss with your members how to improve the appearance or organize the positioning of concepts. You can remove or add more concepts. Try reading each linked concept to ensure meaningfulness of the connection made. Different groups of concepts could be colored to improve clarity of organization.

Stage 6: Finalizing Your Concept Map

Come to an agreement as to the final arrangement of concepts that best conveys your understanding of the topic. Convert your map into a permanent presentation for all to view. Be as creative as you can. Use different colors to differentiate subgroups and add pictures to improve understanding.

CONSTRUCTING A CRITERION MAP

A criterion map can be used as an "agreed-upon organization" that best reflects the structure of content domain. According to Ruiz-Primo, Shultz, Li, & Shavelson (1998), maps drawn by "experts" in the domain (e.g., teachers) provide a reasonable representaion of the subject domain. The suggested steps taken in constructing a criterion map follows:

1. Select a panel: experts in content domain, teachers, researchers, or assessors.
2. Each expert provides a list of the most important concepts in the subject domain.
3. Experts compare and contrast their selected lists and reach a consensus about which are the most important; this forms the "Key-Concept List."
4. Each participant constructs a map using Key-Concept List.
5. Construct a concept map with relations that appear in at least 80% of the participants' concept maps.
6. Discuss and modify resulting concepts.
7. Use resulting concept map as a Criterion Map. (Ruiz-Primo et al., 1998)

Features of a Good Map

Some of the features of a good concept map include

1. A concept map drawn need not be symmetrical in appearance. A good concept map is not just a group of concepts linked to each other but shows which concepts are more important by their placement on the map and what concepts branch off them.
2. A good concept map does not have more than three words in a concept box or more than three boxes on one level. A labeled line should link each pair of concepts. Labels should not be lengthy and should make the connections between two concepts clear. Examples should not be boxed.
3. There are no perfectly correct concept maps; each individual will display a map closest to his/her understanding of the subject. This makes concept mapping unique in portraying students' understanding.

Concept Mapping Techniques

There are several ways we can approach concept mapping. One of the factors that teachers should consider is the amount of directedness given in concept mapping. Teachers can approach concept mapping with little directedness where students are given only the seed concept while all other concepts related to it would come from the students. A second approach would be to provide students with a seed concept and a list of related concepts. Students are required to develop their own concept maps. An example of this approach is shown in Figure 6.4.

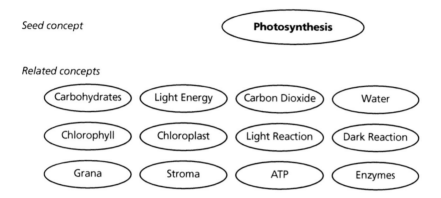

Figure 6.4 Drawing a concept map with the seed and related concepts given.

Another way of approaching concept mapping would be concept maps that are incomplete. Maps given may have incomplete nodes or incomplete links. These maps are called Fill-in-the Nodes or Fill-in-the Links maps, respectively. Examples of these maps are given in Figures 6.5 and 6.6.

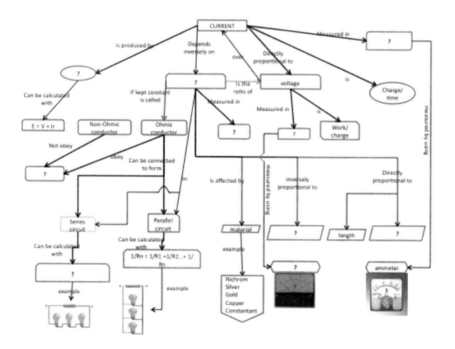

Figure 6.5 A "Current" concept map with some missing nodes.

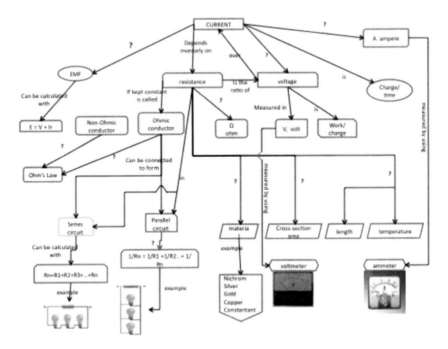

Figure 6.6 A "Current" concept map with some missing links.

Assessing through Concept Maps

Concept maps can also be used as an alternative form of assessment. Within a given concept map are various characteristics of authentic assessments. Joseph Novak (1990) of Cornell found that an important by-product of concept mapping is its ability to detect or illustrate the misconceptions learners may have in a particular subject matter. Concept maps drawn by students express their conceptions (or misconceptions) and can help the instructor diagnose the misconceptions. A study by Novak, Gowin, and Johansen (1983) suggests that concept maps tap into substantially different dimensions of learning than conventional classroom assessment techniques.

On the use of concept maps to assess science understanding, Mintzes, Wandersee, and Novak (2000) stress that concept maps provide teachers with an avenue for developing insight into students' understanding, as evidenced by well-organized and richly elaborated knowledge structures and valid propositional relationships and interrelationships. Concept maps also help to identify errors, omissions, or misunderstanding; and they depict the important organizational function certain concepts play in shaping understanding (Mintzes et al., 2000). Based on their belief that an assessment should be based on a task, a response format, and a scoring system, Ruiz-Primo and Shavelson (1996) proposed a framework for using concept maps as a potential assessment tool in science. A concept map used as an assessment tool should therefore include (a) a task that invites students to provide evidence on knowledge in a certain domain, (b) a format for the students' response, and (c) a scoring system by which the students' concept maps can be accurately and consistently evaluated (Ruiz-Primo & Shavelson, 1996).

There are several methods by which concept maps can be assessed:

Novak & Gowin's Scoring System (1984)

Novak and Gowin evaluate concept maps on the basis of having levels of hierarchy, the validity of propositions and cross-links, and use of examples. Table 6.2 illustrates this scoring system.

Markham et al. (1994) Scoring System

This scoring system scored six observed aspects of students' maps: (a) the number of concepts, as evident of extent of domain knowledge; (b) concept relationships, which provide additional evidence of the extent of content knowledge; (c) branching, which they viewed as evidence of progressive differentiation; (d) hierarchies, providing evidence of knowledge subsumption; (e) cross-links, which represent evidence of knowledge inte-

TABLE 6.2 Novak and Gowins Scoring System

Component	Description	Score
Proposition	Is the meaning relation between two concepts indicated by the connecting line and linking word(s)? Is the relation valid.	1 point for each meaningful, valid proposition shown
Hierarchy	Does the map show hierarchy? Is each subordinate concept more specific and less general than the concept drawn above it?	5 points for each level of hierarchy
Crosslinks	Does the map show meaningful connections between one segment of the concept hierchy and another segment?	10 points for each valid and significant cross link
Examples	Specific events or objects that are valid instances of those designated by the concept level.	1 point for each example

gration; and (f) examples, which indicate the specificity of domain knowledge. Table 6.3 illustrates this scoring system.

Another approach to assessing a concept map is to compare students' concept maps with a criterion map and score the overlap between them.

TABLE 6.3   Concept Map Scoring System (Markham, et al., 1994)

Component	Description	Score
Concepts	No. of concepts	1 point for each concept
Concept relationships	No. of valid relationships	1 point for each valid relationship
Branching	Scores for branching varied according to the amount of elaboration	1 point for each brancing 3 points for each successive branching
Hierarchies	No. of hierarchies	5 points for each level of hierarchy
Cross-links	No. of cross-links	10 points for each cross-link
Examples	No. of examples	1 point for each example

Source: McClure et al. (1999)

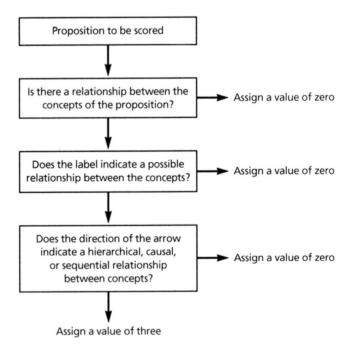

Figure 6.7   Scoring system for a concept network proposed by McClure et al. (1999).

Lomask, Baron, Greig, & Harrison (1992) proposed a method of scoring based on the count of terms and links. The size of the count of terms was expressed as a proportion of terms in an expert concept map mentioned by the students. This proportion was scaled from complete (100%), to substantial (67%–99%), to partial (33%–66%), to small (0%–32%), to none (no terms mentioned or irrelevant terms mentioned). They characterize the strength of links between concepts ranging from strong (100%) to medium (50%–99%), to weak (1%–49%), to none (0%). Table 6.4 shows the scores that took into account both the size of terms and strength of links as suggested by Lomask et al.

Concept maps can also be assessed qualitatively by analyzing the quality of propositions developed (Table 6.5).

TABLE 6.4 Scoring System Based on Size and Strength of Students' Concept Maps

Size	Strength			
	Strong	**Medium**	**Weak**	**None**
Complete	5	4	3	2
Substancial	4	3	2	1
Partial	3	2	1	1
Small	2	1	1	1
None/Irrelevant	1	1	1	1

TABLE 6.5 Qualitative Analysis of Concept Maps Based on the Quality of Propositions

Quality of proposition	Definition	Points
Excellent	Outstanding proposition, complete and correct. Shows deep understanding of the relation between two concepts.	4
Good	Complete and correct proposition. Shows good understanding of the relationship between two concepts.	3
Poor	Incomplete but correct proposition	2
Don't care	Incomplete but correct proposition	1
Invalid	Incorrect proposition	0

COMPUTER-AIDED CONCEPT MAPPING TOOL

According to Jonassen (1990), concept mapping computer tools belong to a rare category of computer tools that were designed specifically for learning. He listed some of the advantages of computer support for concept mapping:

1. *Ease of adaptation and manipulation*: Concept mapping computer tools encourage revisions to the concept map because deletions, additions, and changes can be achieved quickly and easily.
2. *Dynamic linking*: Most computer-assisted concept mapping applications allow learners to point, drag, and drop a concept or groups of concepts to another segment of the map and automatically update all the relevant links.
3. *Conversions*: Most computer-assisted concept mapping applications allow learners to convert the map into different electronic formats. These can be in the form of images, PDF files, or even a hypertext structure. These formats can be saved, exported, printed, or deleted just like any other computer file.
4. *Communication*: Maps constructed can be easily sent as attachments in e-mails or uploaded onto Web pages for ease of communication with collaborators.
5. *Storage*: The use of computer-assisted concept mapping tools allows for digital storage. This kind of storage takes less space and makes retrieval easier.

Examples of computer-assisted concept mapping tools are shown in Table 6.6.

TABLE 6.6   Some Concept Mapping Computer Tools

Concept Mapping Tool	Description
CMap Tools	• Users can construct, navigate, share, and criticize knowledge models represented as concept maps • Allows user to split large maps into collections of smaller concept maps • Also facilitate the linking of maps, enabling the navigation from one Cmap to another • User can also establish links to other types of resources (images, videos, sound clips, etc.)
Bubbl.us	• Free Web application that lets you brainstorm online • Create colorful mind maps online • Share and work with friends • Embed mind map into blog or Web site • E-mail and print mind maps • Save mind map as image
Webspiration	• Can be used to map out ideas, organize with outlines, and collaborate with teams of learners • Able to transfer to Word and Google docs • Can post comments and questions
MindMeister	• Facilitates real-time collaboration • Create, manage, and share mind maps online • Can simultaneously work on the same map and see each others' changes as they happen
FreeMind	• Free online mind mapping software
DropMind	• Picture your thoughts

REFERENCES

Arnaudin, M. W., & Mintzes, J. J. (1985). Students' alternative conceptions of the human circulatory system: A cross-age study. *Science Education, 69*(5), 721–733.

Ausuble, D. (1963). *The psychology of meaningful learning*. New York, NY: Grune and Stratton.

Chi, M., Glaser, R., & Farr, M. (1988). *The nature of expertise*. Hillsdale, NJ: Erlbaum.

Dochy, F. J. R. C. (1994). Assessment of domain-specific and domain transcending prior-knowledge: Entry assessment and the use of profile analysis. In M. Birenbaum & F. J. R. C. Dochy (Eds.), *Alternatives in assessment of achievements, learning process and prior knowledge* (pp. 93–129). Boston, MA: Kluwer Academic.

Edmonson, K. (2000). Assessing science understanding through concept maps. D. J. Mintzes, J. Wandersee, & J. Novak (Eds.), *Assessing science understanding* (pp. 19–40). San Diego, CA: Academic Press.

Jonassen, D. H. (1990, July). What are cognitive tools? In P. A. M. Kommers, D. H. Jonassen, & J. T. Mayes (Eds.), *Proceedings of the NATA advanced research workshop: Cognitive tools for learning* (pp. 1–6). Enschede, The Netherlands: University of Twente.

Jonassen, D. H. (1996). *Computers in the classroom: Mindtools for critical thinking.* Engelwood Cliffs, NJ: Merrill/Prentice Hall.

Jonassen, D. H. (1997). Instructional design models for well-structured and ill-structured problem-solving learning outcomes. *Educational Technology: Research and Development, 45*(1), 65–95.

Jonassen, D. H., C. Carr, & H-P. Yueh. (1998). Computers as mindstools for engaging learners in critical thinking. *TechTrends, 43*(2), 24–32.

Kinchin, I. M. (2000). Concept mapping in biology. *Journal of Biological Education, 34*(2), 61–68.

Kinchin, I. M., Hay, D., & Adams. (2000). How a qualitative approach to concept map analysis can be used to aid learning by illustrating. *Educational Research, 42*(1), 43–57.

Kinchin, I., & Hay, D. (2005). Using concept maps to optimize the composition of collaborative student groups: A pilot study. *Journal of Advance Nursing, 51*(2), 182–187.

Lomask, M., Baron, J. B., Greig, J., & Harrison, C. (1992). *ConnMap: Connecticut's use of concept mapping to assess the structure of students' knowledge of science.* Paper presented at the annual meeting of the National Association of Research in Science Teaching, Cambridge, MA. March 21–25.

Markham, K. dan Mintzes, J. J. (1994). The concept map as a research and evaluation tool: Further evidence of validity. *Journal of Research in Science Teaching, 31*(1), 91–101.

McClure, J. R., Sonak, B., & Suen, H. K. (1999). Concept map assessment of classroom learning: Reliability, validity and logistical practicality. *Journal of Research in Science Teaching, 36*(4), 475–492.

Mintzes, J. J., Wandersee, J. H., & Novak, J. D. (2000). *Assessing science understanding: A human constructivist view.* San Diego, CA: Academic Press.

Novak, J. D. (1990). Concept mapping: A useful tool for science education. *Journal of Research in Science Teaching, 27*(10), 937–949.

Novak, J. D., & Gowin, D. B. (1984). *Learning how to learn.* New York: Cambridge University Press.

Novak, J. D., Gowin, D. B., & Johansen, G. T. (1983). The use of concept mapping and knowledge: Vee mapping with junior high school science students. *Science Education, 67*(5), 625–645.

Okada, T., & Simon, H. A. (1997). Collaborative discovery in scientific domain. *Cognitive Science, 21*(2), 109–104.

Osmundson, E., Chung, G. K. W., Herl, H. E., & Klein, D. C. D. (1999). *Concept mapping in the classroom: A toll for examining the development of students' conceptual understandings* (CSE Tech. Rep. No. 507). Los Angeles: University of California, National Center for Research on Evaluation, Standards, and Student Testing (CRESST).

Ruiz-Primo, M. A., & Shavelson, R. J. (1996). Problems and issues in the use of concept maps in science assessment. *Journal of Research in Science Teaching, 33*(6), 569–600.

Ruiz-Primo, M. A., Shavelson, R. J., & Shultz, S. E. (1997). *On the validity of concept-map based assessment interpretations: An experiment testing the assumption of hierarchical concept maps in science* (CSE Technical Report No. 455). Los Angeles: UCLA-CRESST.

Ruiz-Primo, M. A., Shultz, S. E., Li, M., & Shavelson, R. J. (1998). *Comparison of the reliability and validity of scores from two concept-mapping techniques* (CSE Technical Report No. 492). Los Angeles: UCLA-CRESST.

Shavelson, R. J., Baxter, G. P., & Gao, X. (1993). Sampling variability of performance assessments. *Journal of Educational Measurement, 30,* 215–232.

Shavelson, R. J., & Ruiz-Primo, M. A. (1998). *On the assessment of science achievement: Conceptual underpinnings for the design of performance assessments: Report of year 2 activities.* (CSE Technical Report No. 491). Los Angeles: UCLA-CRESST.

Stevens, R. H., Lopo, A. C., & Wang, P. (1996). Artificial neural networks can distinguish novice and experts strategies during complex problem-solving. *Journal of the American Medical Informatics Associations, 3,* 131–138.

CHAPTER 7

PHYSICS MODELING

An Approach Stimulates
Students' Conceptual Understanding
of Scientific Concepts

Funda Ornek
Bahrain Teachers College, University of Bahrain

I *hear and I forget*
see and I remember
do and I understand
—Confucius

ABSTRACT

Models and modeling have had an important role in constructing new knowledge of scientific concepts and developing scientific understanding of science students. This chapter emphasizes physics modeling and how it can promote students' conceptual understanding of scientific concepts and develop and challenge their higher-order thinking in physics. Physics modeling means making idealizations of a complex system, making approximations, simplifying assumptions, and applying fundamental physical principles to explain the behavior of the physical system (Chabay & Sherwood, 1999). The significance

Contemporary Science Teaching Approaches, pages 137–162
Copyright © 2012 by Information Age Publishing
All rights of reproduction in any form reserved.

of physics modeling is that it provides authentic representation of contemporary physics and encourages students to construct their own knowledge through exploring some aspect of physics concepts by using fundamental principles rather than given formulas. This chapter also emphasizes some modeling activities and relates some experiences from implementation of the physics modeling approach in teaching physics.

INTRODUCTION

Models and modeling are central aspects of constructing new knowledge, developing scientific understanding of physical phenomena, and doing science. Models and modeling were also recommended as a unifying theme for science education by the National Science Education Standards (NRC, 1996). Broadly, modeling corrects many flaws of conventional instruction-demonstrations, lab activities, and tutorials, including the fragmentation of knowledge such as phenomenological primitives (p-prims), which are knowledge in pieces (diSessa, 1993), the passive role of students, and the resistance of changing naïve beliefs about the physical world. As known, there are different types of modeling in science education, such as mathematical modeling, physical modeling, computer modeling, physics modeling, and so forth. This chapter focuses on physics modeling and how it can promote students' conceptual understanding of scientific concepts in order to challenge their higher-order thinking in physics and overcome misconceptions that are related to promoting conceptual change. Physics modeling has become an important constructivist pedagogy in physics teaching as well as in constructing new knowledge of scientific concepts. Physics modeling stimulates and supports students to engage in the process of modeling the real-messy complex world by making simplifications, idealizations, approximations, and testing their models by comparing their results to the real-world system and revising their modeling if required.

PHYSICS MODELING

As is well known, the goal of contemporary physics is to elucidate a wide range of physical phenomena surrounding us or in space. For instance, when you take off your sweater, you can generate a voltage of approximately 30,000 volts. This electricity cannot be utilized because a spark lasts only a short time. We would like to know why a spark does not last a long time and why we cannot make use of the electricity. The reason will be explained further in this chapter by applying physics modeling. Nevertheless, this goal is not given to students who take physics courses in conventional physics instruction. Students think that there should be specific equations to solve specific problems, and their misunderstanding leads them to ask which formula to use in problem solutions. Students can obtain right answers without thinking, understanding, or analyzing the system, or by making small changes to some problems solved in the past, or by using the plug-and-chug method. Accordingly, the emphasis is mostly on solving problems by using certain formulas with the plug-and-chug technique rather than on analyzing every situation by applying the fundamental principles. Plug and chug means finding a specific formula to solve a given problem, plugging the numbers into the formula, and getting another number as an answer, even if you do not understand the concepts behind the formula or what the answer you obtain means.

A contemporary physics curriculum based on physics modeling has been developed by Chabay & Sherwood (1999) to enhance students' conceptual understanding of a broad range of physical phenomena, starting from the fundamental principles to analysis of the physical system. Moreover, physics modeling provides an opportunity for students to engage in the process of structuring their models and analyzing new physical phenomena that are different from the ones they have already seen.

The difference between physical modeling and physics modeling may not be clear at first. Physical models are considered as models of real situations and can be carried, touched, or held (Ornek, 2008). They are a physical representation of the system or an object. The system or the object can be small or large. For example, a physical modeling of an atom can be small or large; it can be larger than a physical modeling of our solar system, although the original is far greater in dimension. See, for example, Figure 7.1 from my

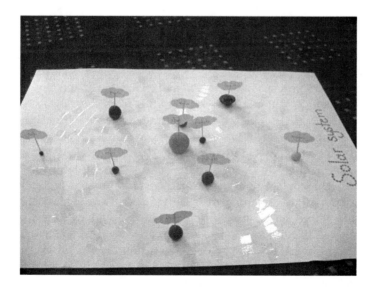

Figure 7.1   A physical model of the solar system.

students' work. Physical models can be used in several fields, such as physics, chemistry, biology, architecture, or engineering, etc., whereas physics modeling is specific for physics, and it always starts from powerful fundamental principles. Of course, both have some similarities such as simplifications, assumptions, approximations, and idealizations. Although Dr. Ruth Chabay and Dr. Bruce Sherwood, who are the designers of physics modeling, use the notation of physical modeling instead, I prefer not to use this designation. Instead, I prefer physics modeling because it is completely related to physics, and it gets confused easily with physical modeling by students.

Physics modeling is the process of idealizing a real-messy complex system, making idealizations, explicit approximations, simplifying assumptions, and applying fundamental principles such as the linear momentum principle or the energy principle to explain the behavior of a complex system (Chabay & Sherwood, 2004). These fundamental principles are in mechanics, electricity, and magnetism. In mechanics, the linear momentum principle, the energy principle, and the angular momentum principle are employed to start physics modeling. In electricity and magnetism, the conservation of charge and the field concept are added to the principles used in mechanics. In this chapter, the emphasis and examples about physics modeling will be mostly from mechanics. In other words, the fundamental principles in mechanics are explained and used in several examples in this chapter.

The advantage of physics modeling is that students are active in the process of making models by applying fundamentally powerful principles including simplifications, idealizations, and approximations instead of using modeling done by the teacher, instructor, authors of books, or professors. Even some physics students, both secondary and undergraduate level, believe that the physics they learn in introductory physics courses is not related to real life. They see it as a "vacuumed" subject, because everything is idealized for them. For example, they see a string as massless, a pulley as having a negligible mass, and friction in a pulley system and environment as being airless. All these idealizations, simplifications, estimations, approximations, and modeling have already been done for them. So students do not know the reasons why we assume that string is massless, and so on (this is a relative approximation with respect to mass and friction involved with other elements, which are always strings and blocks (loads). Pulleys are used in different combination with these elements). We can make up an analogy to make it clear why it is important that students participate in the process of making modeling. Let's say you need to go to one place in your city where you have never been. It is not within walkable distance so you need to drive yourself. You have only two options: you can ask your friend or someone who knows the place to take you there with his/her car, or you can ask directions and drive there yourself. Which one do you think is more constructive in terms of knowing how to get there for future visits? You may say both ways are fine. But in general, the latter one is better than the first one because you are involved in following the directions and finding the place by yourself. You will remember the path you took. Therefore, students can construct new scientific knowledge of physics concepts without memorizing or using a formula provided for them. This process of knowledge construction can assist students to develop their higher-order thinking skills, because higher-order thinking includes complex judgmental skills, critical thinking, problem solving, analyzing, and synthesizing. Physics modeling can help them use their higher-order thinking in novel situations rather than employing rules, formulas, facts, and mathematics through repetitive routines given to them. As a result, the process allows students to make models to solve problems and discover new understandings of physical phenomena. They will be producers of knowledge as they engage in higher thinking by participating in the process of physics modeling. Here is an example of a student's reflection from Ornek's study (2010) with regard to their engagement in the process of making physics modeling:

> The whole concept of modeling that we're not—like in my other physics classes it would be like "here's the equation, this is—this is not a model—this is. This is how this behavior works." And we would neglect air resistance and friction, we would be like they're negligible, whereas in this class, that it's a model and we're purposely leaving things out to simplify it kind of thing.

In physics modeling (Chabay & Sherwood, 1999), the following process is followed:

- Start from fundamental principles
- Estimate quantities
- Make assumptions and approximations
- Decide how to model the system
- Explain/predict a real physical phenomenon in the system
- Evaluate the explanation or prediction

In summary, physics modeling has a potential to provide students with an opportunity to attempt to explain a broad range of physical phenomena rather than having all idealization, simplification, approximation, estimation, and modeling already done for them. It provides students with a chance to engage in the process of constructing models by means of simplifying; idealizing messy, complex, real-world systems; making conscious approximations and estimations (Chabay & Sherwood, 2004) themselves. Physics modeling allows students to use only a few fundamental principles, such as the linear momentum principle, the energy principle, and the angular momentum principle to analyze the systems. They can employ one or more of these principles to analyze several systems. In other words, they do not need to memorize all equations to solve the problems and analyze the systems. On the other hand, students need to use algebra and calculus to be able to model the complex messy systems.

Students' own reflections are one way of finding out how they experience and understand the power of fundamental principles and application of the principles. The other way to assess students' understanding is to examine their problem-solving ability by using the fundamental principles to solve physics problems. The latter is explored later in this chapter. Students who were in an introductory calculus-based physics course in which physics modeling was employed were asked to write their thoughts about the way in which the course focused on modeling the behavior of the real physical world. The following quotations are examples of different students' reflections from Ornek's study (Ornek, 2010).

Students' Reflections on Physics Modeling

There's more to just physics than memorize this humongous block of equations the teacher says works. Uh, and plugging in the numbers and knowing how to put the equations together. We're going back and, well, we can, we're really creating a model of this system and we can, you know, get rid of these factors because the gravitational pull here is really not going to effect how me jumping off a chair is going to do anything. So what we can neglect even though there really is a force there it's small enough that we don't have to. Just kind of learning about physics in a very organized, and going back to elementary steps manner.

It [modeling teaching approach] can be a little frustrating at times because in high school, we didn't go all the way back to where we had to base everything off the momentum principle kind of thing. So it's a little frustrating to look at a problem and be like "I've done this problem before. I know exactly how to solve it the way I've learned before. And now I've got to try and figure out how to solve it and prove that I know how to solve it using a new method." It's kind of cool because the momentum principle applies to just about anything, so you don't have to worry about memorizing this block of equations because you can always re-derive it from three or four fundamental equations we use in class.

I liked that we were taught and expected to know the underlying concepts of the equations and principles. Modeling made me really understand the physics, rather than the mass. Like, I learned how things work.

Powerful Fundamental Principles

In physics modeling, students use a few powerful fundamental principles that make it possible to explain a very large range of physical phenomena in physics. Students are expected to be capable of employing these principles in different situations. As mentioned above, only three powerful principles in mechanics are deeply examined in this chapter. Constructing a physics model always starts from the fundamental principles. Some of the principles that are utilized in mechanics follow:

The Linear Momentum Principle (Chabay & Sherwood, 2002a): A change of momentum is equal to net force times duration of interaction. This principle is also known as Newton's second law.

$$\Delta \vec{P} = \vec{F}_{net} * \Delta t \Rightarrow \text{as } \Delta t \rightarrow 0 \text{ (for small } \Delta t), \quad \frac{\Delta \vec{P}}{\Delta t} \rightarrow \frac{d\vec{P}}{dt} = \vec{F}_{net}$$

$\Delta \vec{P}$: Change of magnitude and direction of linear momentum

$\vec{F}_{net} * \Delta t$: Impulse (the product of force times time)

So, Change in momentum equals to net impulse due to the net force.

Momentum is defined relativistically as

$$\vec{P} = \frac{m\vec{v}}{\sqrt{1 - \dfrac{v^2}{c^2}}} \quad (m \text{ is the mass at rest}).$$

When we make this approximation, $v \ll c$ because no matter how long we keep applying force, and no matter how large the linear momentum becomes, the speed never gets as big as c, which is a cosmic speed limit, thus we will obtain nonrelativistic momentum which is $\vec{P} \approx m\vec{v}$. If we substitute this approximated nonrelativistic momentum in the linear momentum principle, we will obtain the following, which gives us *Newton's second law.* Isn't it familiar to you?

$$\frac{d\vec{P}}{dt} = \frac{d(m\vec{v})}{dt} = m\frac{d\vec{v}}{dt} = \vec{F}_{net} \Rightarrow \vec{F} = m\vec{a}$$

I am sure it is. Physics instructors, teachers, and students all know Newton's second law by heart. However, I remember physics courses in the traditional curriculum that I took in my undergraduate program (Engineering Physics) in the early 1990s. We were not taught in a way that allowed us to obtain Newton's second law and others from the linear momentum principle. What we did, as many students have probably done, was to memorize all equations to be able to solve physics problems on exams. We had no opportunity to experience the power of the Newtonian synthesis, in which motion is predicted from initial conditions and a force law. Therefore, in the traditional physics courses, we did not have any idea that a small number of powerful fundamental principles could explain a very large range of physical phenomena but found instead a formula for each situation by using the plug-and-chug method.

Now let's look at how we can obtain Newton's first and third laws from the linear momentum principle.

$$\text{Newton's First Law: } \frac{d\vec{p}}{dt} = \frac{d(m\vec{v})}{dt} = m\frac{d\vec{v}}{dt} = \vec{F}_{net}$$

since it will be either constant velocity or no velocity.

If velocity is constant or uniform,

$$m\frac{d(v)}{dt} = m0 = 0 \Rightarrow F_{net} = 0$$

If velocity is 0,

$$m\frac{d(0)}{dt} = m0 = 0 \Rightarrow F_{net} = 0$$

Hence, we found total force is 0 (zero) in both cases. We can conclude that there is no interaction, so there is no change in an object's motion. In other words, if an object is moving in a straight line, it will continue to move with no change of direction and magnitude of speed unless there is an interaction. If it is not moving, it will stay at rest. That means the speed of the object is zero.

Newton's Third Law: Let's start to demonstrate the third law with Figure 7.2.

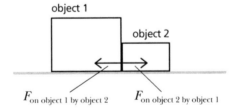

Figure 7.2 The object 1 and the object 2 apply equal and opposite forces on each other.

Since there are no external forces (isolated system), change in the momentum will be zero. So

$$\frac{d\vec{P}_T}{dt} = \vec{F}_{net} = 0 \Rightarrow \frac{d\vec{P}_{12}}{dt} + \frac{d\vec{P}_{21}}{dt} = 0 \Rightarrow \frac{d\vec{P}_{12}}{dt} = -\frac{d\vec{P}_{21}}{dt} \Rightarrow \vec{F}_{on\ 1\ by\ 2} = -\vec{F}_{on\ 2\ by\ 1}$$

Thus, we can obtain Newton's third law by using the linear momentum principle. It is also known as reciprocity. According to Newton's third law, using the above figure, the force that object 1 exerts on object 2 is equal and opposite to the force that object 2 exerts on object 1, as seen in Figure 7.2. Moreover, the linear momentum principle can assist students to overcome intuitive or naïve ideas about physics. In a more traditional curriculum, students, in general, have p-prim on Newton's third law when a system has different masses and retain p-prim of a basic concept of Newton's third law. For instance, if one car and truck have a head-on collision, most students think that the truck will exert a bigger force than the car because the truck has larger mass. In addition, playing a tug-of-war is a good example to demonstrate how students have naïve ideas about scientific concepts. Students think that one team wins because of different masses of both teams. This will be further discussed throughout this chapter.

The Energy Principle (Chabay & Sherwood, 2002a): A change in a particle's energy is the final energy minus the initial energy $(E_{final} - E_{initial})$. This change is equal to work (W) done by a net force $\vec{F}$, and is represented $\Delta E = W$. That means if we do work on a system, we change its energy. This is only for one particle in the system. The energy principle for a multiparticle system is $\Delta E_{system} = W_{by\ external\ forces} \Rightarrow E_{system} = (E_1 + E_2 + ...) + U$, where E_1, E_2 are the particles' energies in the system. U is the potential energy of interacting particles in the system. The important difference between the particle work-energy relation and the multiparticle energy principle is the potential energy U related to interactions inside the system.

Let me give an example with regard to the energy principle and how to apply it to solve problems.

(Example adapted from Chabay & Sherwood, 2002a, p. 125): Let's say you are pulling a 4 kg block as shown in Figure 7.3 and change its speed from 5 m/s to 6 m/s. What is the minimum change of chemical (food) energy in you? Please give magnitude, sign, and units too.

Solution:

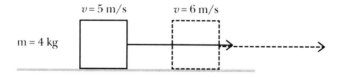

Figure 7.3 Pulling a block.

Approximations made: We assumed that the speed is quite small compared to c, so we will use a nonrelativistic work-energy relation.

Your body temperature will increase when you do work like this because you are moving a block from one point to another point. And thermal energy is also increasing.

The change in kinetic energy is

$$\Delta K = \frac{1}{2}mv_f^2 - \frac{1}{2}mv_i^2 = \frac{1}{2}m(v_f^2 - v_i^2) = \frac{1}{2}(4kg)\left[(6m/s)^2 - (5m/s)^2\right]$$

$$= 2(36-25) = 2(11)$$

$$\Delta K = 22 \text{ Joule}$$

From the Energy principle, since it is not a multiparticle system (a particle), we can use $\Delta K = W$.

$\Delta K = 22J$ so your work done by pulling the block is equal to +22J. It means your store of chemical (food) energy decreased 22J. So the change is −22J.

In a traditional curriculum, we would not be involved in the process of making approximations. We would not even think about relativistic work-energy relations, nor be aware of them. If relativity is not introduced in the beginning of a mechanics course, students will not be aware of relativistic relations. The most important thing is that they get confused when it comes to Einstein's world because students start to feel that whatever they are taught in Newton's world is not correct in Einstein's world. Whereas if they were in the process of making models by approximations, implications, and idealizations, they would not be confused and think that physics is inaccurate.

The Angular Momentum Principle (Chabay & Sherwood, 2002a): The rate of change of the angular momentum of a particle relative to a location is equal to the torque applied to the particle about that location. This is

$$\frac{d\vec{L}}{dt} = \vec{r} \times \vec{F}_{net} = \vec{\tau}.$$

The angular momentum principle for a multiparticle system is

$$\frac{d\vec{L}_{tot}}{dt} = \vec{\tau}_{net,external}$$

which is the rate of change of the total angular momentum of a system relative to a location, $\vec{L}_{tot} = \vec{L}_1 + \vec{L}_2 + \vec{L}_3 + ...$, is equal to the net torque due to external forces exerted on that system relative to the location.

(Example adapted from Chabay & Sherwood, 2002a), p. 320): Let's think about this scenario. You took your children to a park and let them play. They started to play on a seesaw and called you to watch them while they were playing, as in Figure 7.4. Immediately you started to analyze the system as a physicist as follows:

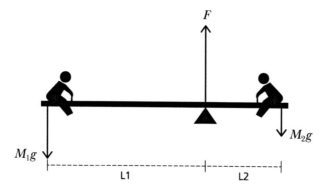

Figure 7.4 Children play on a seesaw.

Approximations: The seesaw is not rotating around its frictionless support. Therefore there is no change in the angular momentum. That means the change in the angular momentum is equal to zero. The board has negligible mass when compared to your children sitting on it.

So you started to write a force equation and a torque equation around the fulcrum (support) to analyze the system by applying the linear momentum and the angular momentum principles

$$\frac{dP}{dt} = F_{net} = F - (M_1g + M_2g) = F - M_1g - M_2g = 0$$

$$F = M_1g + M_2g \text{ Newton}$$

$$\frac{dL_{support}}{dt} = \tau_{net} = L_1M_1g - L_2M_2g = 0 \Rightarrow L_1M_1g = L_2M_2g \Rightarrow L_1M_1 = L_2M_2$$

Now you wanted to analyze the system around a location fixed in a space where your older child on the left was sitting.

$$\frac{dL_{left\ person}}{dt} = L_1F - L_1M_2g - L_2M_2g = 0$$

$$L_1F = (L_1 + L_2)M_2g$$

$$F = (M_1 + M_2)g$$

$$L_1(M_1 + M_2)g = M_2g(L_1 + L_2) \Rightarrow L_1M_1g + L_1M_2g = L_1M_2g + L_2M_2g$$

$$L_1M_1g - L_2M_2g = 0$$

You showed that the above equation is equal to the torque equation around the axle, when we take the force equation into consideration. In addition, you showed that it is useful to choose a fixed location such as the fulcrum so that some forces do not create torque around that location. Thus, they are not parts of the angular momentum principle.

As seen, special-purpose formulas are not required for system analysis, with a few exceptions. Students can use fundamental principles and reason from fundamental principles. For example, students can obtain Newton's third law from the linear momentum principle as shown above. These types of problems and students' participation in making approximations, simplification, idealizations, and assumptions are not available in traditional physics curricula. Introductory physics-mechanics textbooks do not include either one.

Fundamental principles are applied in electricity and magnetism as follows:

The Conservation of Charge Principle (Chabay & Sherwood, 2002b): According to the conservation of charge principle, electric charge cannot be created or destroyed. In other words, if the net charge of an object changes, then the net charge of its surroundings should change, but to the opposite sign. For example, as shown in Figure 7.5, when you rub a balloon with a piece of animal fur (it also works with your hair), the atoms of rubber (balloon) pull electrons from the atoms of animal fur (or from your hair), but at the same time your hair or animal fur acquires an equal amount of positive charge. Therefore, the net charge of the object and its surroundings cannot change, so they are always conserved.

Students in general have misconceptions with regard to this principle because they think that the system is left with an imbalance of charge; it is not correct that the object and the surroundings are separately left with an imbalance of charge. The net charge of the system is not changed. For

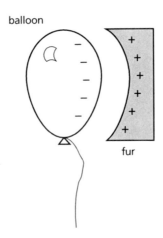

Figure 7.5 Rubbing balloon with a piece of animal fur.

example, I taught a physical sciences course for Math & Science teacher candidates who will teach at the primary level. My students conducted the above activity in the class, and they stuck the charged balloon to the wall and saw that there was an attraction between the wall and balloon (polarization was studied as well). When I asked what happened to the net charge of the system composed of the balloon and a piece of animal fur, some of them gave incorrect answers saying that the system lost charges, whereas the net charge of the system was conserved.

The Concept of Field (Chabay & Sherwood, 2002b): As for the concept of field, there are two concepts of field that can be used in physics modeling. One is electric field and the other one is magnetic field. These are briefly explained in the following section.

Electric Field

Here we will only consider an electric field for a point particle charge not in other situations such as uniformly charged spheres or uniformly charged thin rods. A point charge, let's say, q1 at a location, makes an electric field throughout space. Location is x_1, y_1, z_1. q2 is a different charge at a different location. Let's say, x_2, y_2, z_2 encounters a force because of the electric field that q1 created. $\vec{F_2} = q_2\vec{E_1}$ (Figure 7.6).

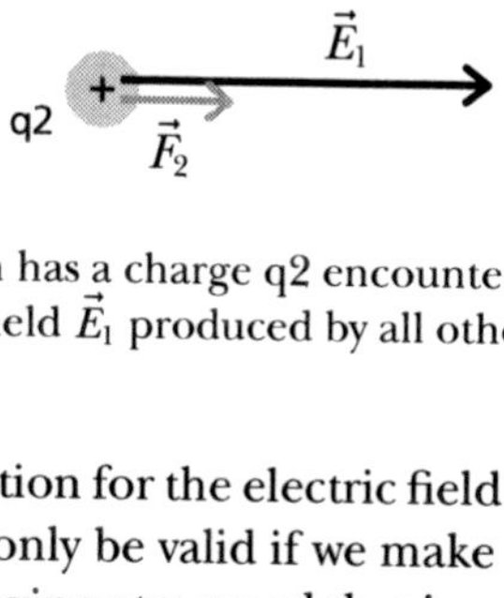

Figure 7.6 A particle which has a charge q2 encounters a force $\vec{F_2}$ due to its interactions with the electric field $\vec{E_1}$ produced by all other charged particles in the neighborhood.

Furthermore, the equation for the electric field of a point particle charge is only obtained and can only be valid if we make a nonrelativistic approximation if the charge is moving at a speed that is very small compared to the speed of light. The relation between an electric field and a charge is called "Coulomb's Law" (Figure 7.7).

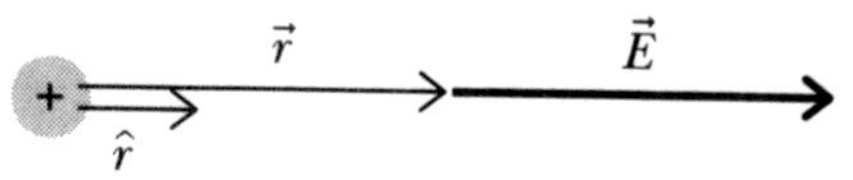

$$\vec{E} = \frac{1}{4\pi\varepsilon_0}\frac{q}{r^2}\hat{r}$$

Figure 7.7 The vector $\vec{r}$ points from the point charge to the observation location. $\hat{r}$ is the unit vector.

In addition, this concept of field includes other principles such as the superposition principle. But these are not discussed in this chapter.

Magnetic Field

Here again, we will only consider a magnetic field for a moving particle, not for other circumstances such as the magnetic field of a wire, magnetic field of a loop, or magnetic field inside a long solenoid. Magnetic fields are produced by moving charges, unlike an electric field, which can be produced by both moving charges and charges at rest. A magnetic field curls around the charge. Moreover, the Biot-Savart Law is a fundamental principle of the concept of a magnetic field.

The Biot-Savart Law for a single particle charge is

$$\vec{B} = \frac{\mu_0}{4\pi} \frac{q\vec{v} \times \hat{r}}{r^2}.$$

$\vec{v}$ is the velocity of the point charge q, and $\hat{r}$ is the unit vector. The Biot-Savart Law cannot be relativistically correct like Coulomb's Law. It can give accurate results if the speed of the moving particle charge are small compared to the speed of the light. So we use approximation to apply the Biot-Savart Law in a nonrelativistic system (Figure 7.8).

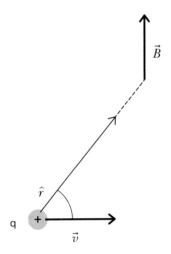

Figure 7.8 The magnetic field produced by a moving particle charge is perpendicular to $\hat{r}$ and $\vec{v}$.

Physics Modeling and Overcoming P-Prims

Students come to classes with intuitive or naïve physics knowledge about collisions, heat, temperature, and energy built based on their experience and interactions with peers, families, and others. The intuitive or naïve ideas about scientific knowledge that students have are retained even if the students complete physics courses in a traditional curriculum, because students are not involved in the process of analyzing the system. Everything is in a closed system and provided to students in isolation. Therefore, the traditional curriculum of physics can lead to students' misunderstanding of physics concepts. Here are two examples from two different courses taught in different countries and different institutions. One course used the physics modeling approach; the other one used mostly hands-on and minds-on activities. Both courses were introductory courses. The following examples are from students' problem solving to show how two different students thought and solved almost the same problem. For both, the instruction medium was English.

Student 1 in country 1

(Question adapted from Ornek's study, 2009a): Imagine that you are driving a Nissan car very happily to BTC. You suddenly realized that a BTC van was coming toward you, and you tried to stop your car, but unfortunately, there was a head-on collision between the BTC van and your Nissan car (Figure 7.9). Each vehicle was initially moving at the same speed. Luckily nothing happened to either driver. You wondered about the nature of forces exerted and started to ask the following questions:

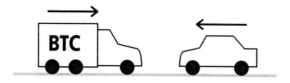

Figure 7.9 Head-on collision between BTC van and a Nissan.

1. Does the BTC van exert (apply) a force on the Nissan?
   Answer 1: Yes
2. Does the Nissan exert (apply) a force on the BTC van?
   Answer 2: Yes
3. If the answers to questions 1 and 2 are yes, which force is larger? Explain your answers to 1, 2, and 3.
   Answer 3: BTC big force because big mass (no reasoning to 1 and 2)

4. Write out a complete statement of Newton's third law of in terms of
forces in *your own words*. (You may use a diagram if you wish).
Answer 4: For every action there is equal and opposite reaction.

Student 2 in country 2 (physics modeling approach was employed in this course)

(Question adapted from Ornek, 2009a). Now think about a head-on
collision between a van and a Ford Escort (Figure 7.10). Each vehicle is
initially moving at the same speed. The following questions refer to what is
happening during collision:

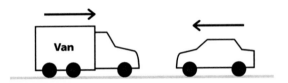

Figure 7.10 A head-on collision between a van and a Ford Escort.

1. Does the moving van exert a force on the Ford Escort?
   Answer 1: Yes, because the Ford Escort has a change in momentum,
   which by the momentum principle, says that there is a force acting
   over a certain period of time. And there's, the only other object in
   this system that it would interact with would be the moving van. So
   the moving van would, is applying the force against the Ford Escort.
   I mean, which the Ford Escort is experiencing.
2. Does the Ford Escort exert a force on the moving van?
   Answer 2: Yes it does. Same thing as the last case. There's a change in
   momentum of the moving van. And other objects being acted, only
   other object being considered in this diagram would be the Ford
   Escort. And the force is a, uh, is proof of an interaction. So any other
   way it could interact with the Ford Escort is through physical contact.
   Which it would be during the collision.
3. If the answers to questions 1 and 2 are yes, which force is larger?
   Explain your answers to 1, 2, and 3.
   Answer 3: The momentum principle of dp equals f net dt. Since
   there is no external forces action on the van and the Ford Escort car,
   the total change of the momentum is equal to zero, so that changes
   the momentum so that it will equal and opposite for each car. dp
   over dt is equal to zero. In addition, the system is nonrelativistic
   model and the mass is not changin. The time interval is also equal

for both the van and the Escort car. So, *dp* over *dt* for the Escort Ford is equal dp over dt for the van

$$\frac{dp_{\text{escort}}}{dt} + \frac{dp_{\text{van}}}{dt} = 0.$$

So

$$\frac{dp_{\text{escort}}}{dt} = -\frac{dp_{\text{van}}}{dt}.$$

As a result, from the momentum principle the force acting on the Escort car by the van is equal and opposite the force acting on the van by the Escort car. ($\vec{F}_{\text{escort car, van}} = -\vec{F}_{\text{van, escort car}}$) which gives us Newton's third law.

4. Write out a complete statement of Newton's third law in terms of forces in *your own words*. (You may use a diagram if you wish).
   Answer 4: Also called the law of reciprocity. Here's a diagram (Figure 7.11).

F12   F21

Figure 7.11 Diagram for showing Newton's third law.

Have two circles labeled one and two force from two pointing to the right when one's on the left. Force on two by one and an arrow the opposite way on one. Force on one by two equals a negative, uh, every force is returned with equal magnitude and opposite direction.

As we compare the first student's answer to the slightly different questions with the second student, we can see how they answered the questions and how their knowledge of physics concepts reflects their problem solving. The striking difference in the two students was that the first student from a more hands-on- and minds-on-based course tried to answer the questions, but could not get the correct answer. They all knew by heart that when the objects have equal masses, the forces acting on each other are equal and opposite to each other. On the other hand, when it comes to different masses, they are not able to figure out what is happening in the system. In this course, all students' answers were not correct in terms of explaining why both cars are exerting forces on each other and which force is larger. They could not even say that the forces acting on each other are equal and opposite because of Newton's third law. This problem was novel for both students because the masses were different. The first student demonstrated intuitive ideas about forces because of different masses because the knowledge s/he had was in pieces. Intuitive physics has fragmented structures that are called p-prims (diSessa, 1993). Intuitive commonsense knowledge obtained from daily-life experiences include isolated and fragmented knowledge pieces, as seen in the first student's answer. Commonsense knowledge is not coherent and consistent (Ornek, 2009b). Therefore, an increase of coherence and consistency should take place to be able to overcome p-prims. The student has a p-prim: "larger mass requires bigger force," which actually is not incorrect in other systems. However, it is not correct when applied to this phenomenon. For example, if you push a block whose mass is 10 kg, you apply less force compared to pushing a block whose mass is 30 kg on the same surface (the coefficient of friction is same) (Figure 7.12). So the p-prim is correct in that context, $F_2 > F_1$.

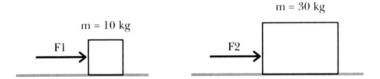

Figure 7.12 Pushing two blocks with different masses.

Therefore, a p-prim can be correct in itself, but its activation may not be correct in other contexts such as in this system. S/he activated the idea that a larger mass requires a bigger exertion than is required to move the small mass in this problem.

Let's look at the second student's performance; he started from the linear momentum principle to solve the problem. In fact, he was unexpectedly successful in solving this problem. He explained how he did his modeling starting from the fundamental principle and given each step he followed. His process was not typical of all students in the course. However, if they had applied the powerful fundamental principles as he did, they would have reasoned correctly and gotten the correct answer. Some slow learners in the second course immediately tried to map the question onto a problem whose solution they already knew in order to solve it. For example, one student said, "Yeah, Newton's third law is, I'm trying to think about it from the class." The most important thing about the physics modeling approach is that students started to think coherently and perform better when they were asked "Why don't you think about the problem in terms of the linear momentum principle?" Suddenly it made sense and students started from a powerful fundamental principle to analyze the system. Most of them had correct reasoning and answered correctly. Of course, there were still students struggling to make sense, because, as they reported, they had not been exposed to physics modeling in their physics courses in the traditional curriculum.

As a result, it can be said that physics modeling has a potential for students to analyze a system correctly and construct a coherent and consistent knowledge so that they can overcome p-prims. As earlier stated in this chapter, problem solving by using powerful fundamental principles was the second way to measure students' views of physics modeling. How students use their knowledge reflects the importance of physics modeling in the nature of knowledge and organization of knowledge as demonstrated above in a problem-solving protocol by applying physics modeling.

Tug-of-war activity: This example is from one of my classes where I was teaching Newton's laws without using physics modeling. The structure of the course is based on mostly hands-on and minds-on activities, tutorials, and lab activities. The course is for second-year Math & Science teacher candidates. During class, I asked my students to play a tug-of-war game to teach the Newton's third law, as seen in Figure 7.13.

I asked students which team was going to win and to explain their reasoning. They immediately said the team that had has two people would win because they have bigger mass. The reasoning was wrong. They thought that the team with the bigger mass exerted a force on the rope greater

Figure 7.13 Students playing a tug-of-war game. (The picture with students' permission.)

than the force exerted by the other team with a smaller mass. Therefore, the team that had bigger mass would win. As you see, they have same p-prim as the one explained in Figures 7.9 and 7.10. Based on my teaching experience, in general, students always have this p-prim when they need to deal with different masses, as shown in the above examples. Students who employ physics modeling, however, are capable of applying the powerful fundamental principles to new situations or problems such as dealing different masses in the above system. Let's analyze a tug-of-war system by applying physics modeling.

Start with a fundamental principle: The fundamental principle here is the linear momentum principle:

$$\frac{d\vec{P}}{dt} = \vec{F}_{net}.$$

We can make the following approximations:

1. The rope has relatively very small mass compared to the masses of team members at the two ends. So the net force on the rope is zero, and any small difference in force (tension) between the two ends is not great enough to account for the outcome.
2. It is a nonrelativistic situation. The speeds of the teams are very small compared to the speed of light.

By using the linear momentum principle, we can obtain the following result, which says the force on the rope applied by team 2 (has larger mass) is equal and opposite to the force on the rope applied by team 1 (less mass) (Figure 7.14). The force on the rope is also called tension and the notation, T, is also used for the tension.

$$\frac{d\vec{P}_{Team1}}{dt} + \frac{d\vec{P}_{Team2}}{dt} = 0 \Rightarrow \vec{F}_{Team1byTeam2} = -\vec{F}_{Team2byTeam1}$$

$$\vec{F}_{Team1byTeam2} \qquad \vec{F}_{Team2byTeam1}$$

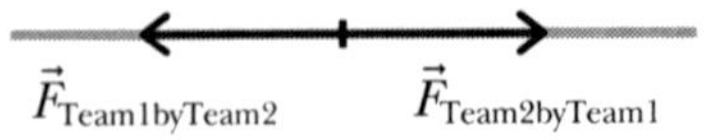

Figure 7.14 Forces on the rope (except forces exerted by the two teams).

Then how are we going to explain why team 2 wins? Again, using the linear momentum principle, we will find the correct answer. Since there is frictional force between team 1 and the floor and team 2 and the floor, there is a contribution from these forces.

First let's analyze the system for team 1 (less mass, m):

$$\frac{d\vec{P}}{dt} = \vec{F}_{net} = F_{team1byrope} - f_{team1}$$

The net force on team 1 that allows team 1 to accelerate with respect to the ground is the force the rope exerts on team 1 minus the force of the ground on their feet (frictional force).

Team 2 (larger mass, M):

$$\frac{d\vec{P}}{dt} = \vec{F}_{net} = f_{team2} - F_{team2byrope}$$

The force on team 1 by the rope is equal to the force on team 2 by the rope because the net force on the rope is zero, as explained above.

The net force on team 2 that allows team 2 to accelerate with respect to the ground is the force of the ground on their feet (frictional force) minus the force of the rope exerted on team 2.

Since the surface is same for both teams, the coefficient of friction is same for both teams. So the frictional force for team 2 will be bigger than the frictional force for team 1. Team 1 also accelerates because the force of the rope on it is larger than the force of the ground on their feet (frictional force). Team 2 wins the game because they can maintain the largest frictional force between their feet and the ground.

Now let's go back to the question "Why does a spark last only a short time when we take off our sweater or when we touch a doorknob?" which we raised in the beginning of this chapter and analyze this situation by using physics modeling.

Why doesn't a spark last a long time? Before we start to explain, we need to make the assumption that the air becomes ionized. How can the air become ionized? We know that an electric field about $E = 3 \times 10^6 \, N/C$ based on the experimental results is enough to ionize air. That means the air becomes a conductor. However, we need to construct a model to compare the predictions of our model to the one obtained experimentally observed (the electric field) to make sure that this model is a good one that we can use to explain why air becomes ionized. We know that an atom or molecule is polarized by an electric field. Therefore, it can be assumed that an electric field of $E = 3 \times 10^6 \, N/C$ is large enough to rip an electron completely out of an air molecule. As a result of this, this phenomenon creates a positive ion and a free electron. Now let's check if our model is good enough to explain why air becomes ionized.

Model 1: A strong electric field rips an electron out of a molecule (Chabay & Sherwood, 2002b). Physics practitioners all know that we need a huge electric field to pull an electron out from an atom. Let's use one of the powerful fundamental principles, the concept of field (here it is electric field concept) to find out the required electric field to pull an electron out from an atom. As known, the radius of an atom is approximately 10^{-10} meter. Moreover, we can assume that a spherical charged object is like a point particle. For the purpose of making calculations simple, it is better to consider a hydrogen atom because it has a positively charged proton and a negatively charged electron—the outer electron is bound to the nucleus because of Coulomb's force. Therefore, the electric field that we need to pull an outer electron out from the hydrogen atom can be found as follows:

$$E = \frac{1}{4\pi\varepsilon_0}\frac{q}{r^2} = (9\times10^9\,Nm^2/C^2)\frac{+e}{(10^{-10}\,m)^2} = (9\times10^9\,Nm^2/C^2)\frac{(1.6\times10^{-19}C)}{(10^{-10}\,m)^2}$$

$$E = 1.4\times10^{11}\,N/C$$

$$\frac{1}{4\pi\varepsilon_0} = 9\times10^9\,Nm^2/C^2;$$

$q = +e = +1.6\times10^9\,Nm^2/C^2$ is the charge of the atom;
$r = 10^{-10}$ m is the approximate value of the radius of the atom;
E = electric field created by +e charge on –e, outer electron.

Based on our model, we found that the electric field needed to pull an electron out of an atom is extremely greater than the one obtained by experimental observation.

$$E = 1.4\times10^{11}\,N/C \gg E_{crit} = 3\times10^6\,N/C$$

In other words, it is not possible to ionize an atom in this model by applying a strong electric field directly. Therefore, we can discard this model, and we need to construct a new model to be able to explain ionization of air. As you may realize, one of the significant strengths that physics modeling, even all kinds of modeling, has is to be able to change or revise your models completely when they do not work. This enables students to engage in the process of producing models to explain a broad range of phenomena.

Model 2: Fast moving particles that are charged knock electrons out of atoms (Chabay & Sherwood, 2002b). To ionize air, we need a considerable force to pull electrons out of air molecules. For this reason, a fast moving charged particle works better than model 1. A fast moving charged particle that collides with an atom in its way is able to rip an electron out from the atom and leave an ionized ion behind it. Another question raised is from where we can supply this fast moving charged particle. Fast moving charged particles are produced by cosmic rays at the top of the atmosphere, and electrons, and alpha particles because of radioactive reactions around us.

To return to the question of why a spark in the air cannot last long enough that we can use this electricity, we assumed that air is ionized and explained how ionization occurs. In order to be able to construct our model to explain the motion of electric charges in ionized air, we need to consider one negatively charged metal ball and one positively charged ball. The air between these two balls is ionized. Since air is ionized, the air contains mobile electrons, positively charged N_2^+ and O_2^+. What will happen to electrons and positively charged ions in the ionized air? As shown in Figure 7.15, electrons move toward the positive ball and the positive ions move toward the negative ball. In this model, electrons accelerate until they collide with a molecule and almost stop moving, and then they start to accelerate again. So there is a kind of "start-stop" motion for each electron.

When electrons collide with the positive metal ball, this phenomenon makes the positively charged ball become less positively charged. If ions collide with the negative metal ball, the negatively charged ball becomes less negatively charged. As a result of these happenings, in both situations, the electric field at all locations in the air between the metal balls will decrease slightly. Therefore, the charge on both the positively and the negatively charged metal balls is decreased. As a result, the decreases in electric field between the balls will not be sufficient to keep the air ionized a longer time. Then the spark will not last long. It will go out quickly.

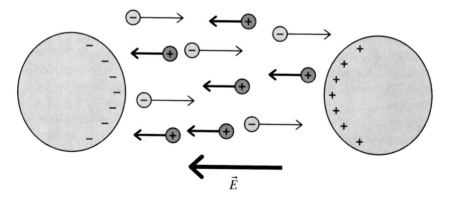

Figure 7.15 Model of ionized air: Movement of mobile electrons and positively charged ions.

CONCLUSIONS

It is important that students understand physical phenomena, construct new knowledge of scientific concepts in physics, and develop their understanding of the scientific concepts in physics, because many contemporary applications of technology and science are based on physics. For example, our society is astonishingly reliant on the findings of physics and on interpretation of these findings. A day without physics would be a day without cars, laptops, cell phones, iPods, or even chocolate cookies and baklava, because it would not be possible to build these without a basic understanding of how the universe works. Understanding how the universe works would never be possible if we relied only on "isolated" physics and solving problems by the plug-and-chug technique. So it would not be possible to explain and analyze a broad range of phenomena around us. Can you imagine trying to design an electric oven without understanding and knowing how electricity travels? Or trying to design an iPod without understanding and knowing what sound is and how sound travels.

As a result, physics modeling propounds authentic depiction of contemporary physics and enhances students' conceptual understanding of scientific concepts in physics so that students can construct new knowledge of scientific concepts in physics or overcome their misconceptions or p-prims. Moreover, physics modeling can foster students' higher-order thinking skills that enable them to think critically (Barak & Dori, 2009) in physics. Physics modeling requires students to be able to analyze, synthesize, reason, comprehend, apply, and evaluate their models. If their models are not reasonable, they need to revise their models. The emphasis of physics modeling is not the drill and practice of problems and activities found in conventional curricula.

In addition, physics modeling provides self-learning, because it requires students to construct their own models by starting from the fundamental principles. It can provide an application of the knowledge to real-world situations. For example, fast moving charged particles are produced by cosmic rays at the top of the atmosphere, and electrons, and alpha particles because of radioactive reactions around us. On the other hand, unfortunately, the phenomenon of knocking out an electron from an atom causes dangerous results. For example, a neutral atom in our DNA can suddenly become an ion and cause genetic mutations and diseases. In addition, this process can affect and damage computer systems.

Students are involved in making connections to other fields such as biology and computer science as seen above. So it is not just an isolated system that integrates different fields with physics. Therefore, it is easy to see how things apply in the real world instead of just looking at equations and dealing with them like in high school. One of the students who took an introductory physics course that uses physics modeling reflects as follows:

> It won't be like high school. In high school, everything is in a closed system and you just need to do mostly plug-and-chug procedure to solve the physics problems. So you need to come in with an open mind and don't assume you know everything about physics, because you don't.

In addition to this, it is also worthwhile to present a quote from a student in Chabay and Sherwood's presentation, which reflects how physics modeling can change their thinking about the physics and the way in which is taught.

> I'm surprised that most physics courses avoid the topics covered in this chapter (nonrigid systems and the energy analysis of systems involving friction) when they can be dealt with as straightforwardly as they are here. Typically friction is described as a "nonconservative force" and left at that.

> I've always realized that most physics courses operate in a dream world of frictionless pulleys and massless springs because many real-world effects can't easily be calculated analytically. This, however, is the first time I've seen a topic which can be dealt with using basic principles and simple algebra (meaning no iterative calculations) but which isn't covered in physics textbooks (at least not in my high school physics textbook).

Physics modeling engages students in contemporary physics endeavors through always emphasizing a powerful number of fundamental principles such as the linear momentum principle, and through engaging students in physics modeling by making approximations, idealizations, simplifications, assumptions, and estimations (Chabay & Sherwood, 1999). In addition, physics modeling avoids simple repetition of high school physics and the drill-and-practice problems usually done in the more traditional physics curriculum. Kuhn (2005) also argues that "the enhancement of higher-order thinking skills is essential for equipping students to participate in and contribute to modern 21st century society." Hence, physics modeling has the potential to involve students in, and allow them to contribute to 21st-century physics enterprises.

ACKNOWLEDGEMENT

I would like to thank Prof Starr Ackley for her constructive feedback.

REFERENCES

Barak, M., & Dori, Y. J. (2009). Enhancing higher order thinking skills among inservice science teachers via embedded assessment [Electronic version]. *Journal of Science Teacher Education, 20*(5), 459–474.

Chabay, R. W., & Sherwood, B. A. (1999). Bringing atoms into first-year physics [Electronic version]. *American Journal of Physics, 67,* 1045–1050.

Chabay, R. W., & Sherwood, B. A. (2002a). *Matter and interactions: Modern mechanics* (Vol I). New York: John Wiley & Sons, Inc.

Chabay, R. W. & Sherwood, B. A. (2002b). *Matter and interactions: Modern mechanics* (Vol II). New York: John Wiley & Sons, Inc.

Chabay, R. W., & Sherwood, B. A. (2004). Modern mechanics [Electronic version]. *American Journal of Physics, 72*(4), 439–445.

di Sessa, A. A. (1993). Toward an epistemological physics [Electronic version]. *Cognition and Instruction, 10,* 105–225.

Kuhn, D. (2005). *Education for thinking.* Cambridge, MA: Harvard University Press.

National Research Council. (1996). *National Science Education Standards.* Washington, DC: The National Academies Press.

Ornek, F. (2008). Models in science education: Applications of models in learning and teaching science. *International Journal of Environmental & Science Education, 3*(2), 35–45.

Ornek, F. (2009a). Problem solving: Physics modeling-based interactive engagement. *Asia-Pacific Forum on Science Learning and Teaching, 10*(2).

Ornek, F. (2009b). An overview of conceptual change models: Some cases from physics. *The International Journal of Creativity & Problem Solving, 19*(2), 61–78.

Ornek, F. (2010). *Modeling-based interactive engagement in an introductory physics course: Students' conceptions and problem solving ability.* Germany: VDM Verlag Dr. Müller.

PART II

USING COMPUTERS IN TEACHING SCIENCE

CHAPTER 8

A FRAMEWORK FOR THE INTEGRATION OF TECHNOLOGY INTO SCIENCE INSTRUCTION

Yilmaz Saglam and Servet Demir
University of Gaziantep, Turkey

ABSTRACT

Today, technology integration has become an important aspect of many professional development programs. If integrated properly, the technological artifacts such as animations, simulations, video clips, or models provide kids with an important visual context for the development of scientific concepts. However, the current studies have indicated that teachers are mostly unwilling to use technology and deficient in the insight into how to teach with technology. The studies also pointed to the fact that in order for an effective integration of technology, teachers need to possess technological pedagogical content knowledge (TPCK). That is, they require knowing what technological artifacts are available for use, which of them are appropriate for the particular purpose, and what sort(s) of instructional strategies and methods could be used for the incorporation of those artifacts into their instruction. The present chapter therefore aims to provide science teachers with a framework for

Contemporary Science Teaching Approaches, pages 165–177
Copyright © 2012 by Information Age Publishing
All rights of reproduction in any form reserved.

a successful integration of technology into their instruction. Furthermore, the information presented in this chapter resulted from a project sponsored by the Scientific and Technological Research Council of Turkey (TUBITAK) grant No. 108K330.

WHY IS THE USE OF TECHNOLOGICAL ARTIFACTS SO IMPORTANT?

To Wertsch (1998, pp. 27–28), learning is a human activity in which there is a dynamic, dialectic, and irreducible tension between agent (human being) and mediational tool (physical or symbolic artifacts). To him, all learning activities are mediated by the artifacts. According to him, pole vaulting, an Olympic sport, can be considered as a mediated action and illustrates well the irreducible tension between agent (the sportsman who vaults) and mediational tool (pole). In this sport, the sportsmen run down a 125-foot runway with a pole in their hands, plant the pole in a small box at the end of the runway, and lift themselves off the ground over a bar. In this activity, the pole by itself could not magically vault over the crossbar and similarly, the agent could not vault over the crossbar without the pole. The skill or mastery in using the pole was indeed developed when those two elements interacted in harmony. Without the materiality of the pole, there would be nothing to act with, and the emergence of the skill could not occur.

In an analogous way, the technological artifacts, which, in this chapter, refers to animations, simulations, video clips, models, and pictures, could be considered as mediational tools for the development of scientific knowledge. For instance, the following image (Figure 8.1) is frequently used in science classrooms for teaching children the concept of the nature of science.

The activity, called fossil footprint, is designed for the grades of 5 to 8. In this activity, the teacher informs students that the footprints belong to prehistoric animals, then consecutively shows the position 1, 2, and 3, and initiates a discussion on the species of animals and what is happening in each position. The students are to observe, interpret, and make inferences from the prints and provide defensible explanations for each position. One of the important outcomes of this activity is to make students realize that scientific knowledge is a human inference or imagination and subject to change as new evidence becomes available. This particular puzzle therefore becomes an important mediational tool in attaining this notion. By manipulating or acting with this tool, the students develop an understanding of how scientific explanations change as new evidence surfaces. Therefore, today, if used properly, the technological artifacts could be important mediational means for the development of scientific knowledge.

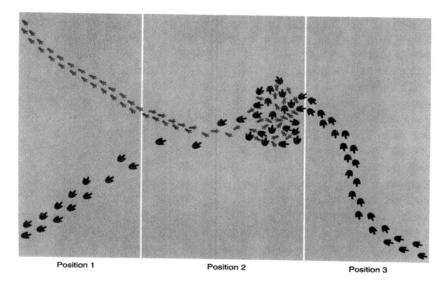

Figure 8.1 Footprint puzzle (National Academy of Sciences, 1998).

A variety of studies in science education indicated that the use of technological artifacts facilitates students' understanding of science (Akpan & Andre, 2000; Barnea & Dori, 1999; Huppert, Lomask, & Lazarowitz, 2002; Kelly & Jones, 2005; Sanger, Phelps, & Fienhold, 2000; Winn, et al., 2006). For instance, according to Bell and Trundle (2008), teaching the concepts of lunar phases with a computer simulation, called Starry Night Backyard, is better than teaching through immediate daily moon observations. To them, the latter method of instruction was time-consuming and frustrating for the students. Furthermore, through that way, the only observation students could make is to see phases of the moon. On the other hand, the animation, Starry Night Backyard, lets the students simultaneously observe the relative position of the Earth, sun, and moon, which thus allows them to realize how the interactions among these bodies cause the emergence of lunar phases.

What Do Teachers Need to Know for Teaching with Technology?

Mishra and Koehler (2005, 2006), inspired by Schulman's (1986, 1987) framework of pedagogical content knowledge, alleged that teachers need to possess technological pedagogical content knowledge (TPCK) in order to teach with technology in an effective way. They took a situated view and believed that TPCK refers to an understanding of the interplay between technology, pedagogy, and content. More recently, Cox and Graham (2009) conducted a study on conceptual analysis of the TPCK framework and provided a precise definition for it. In their study, it was sought to clarify what does and what does not fall under TPCK. They provided various definitions for Pedagogical Knowledge (PK), Content Knowledge (CK), Pedagogical Content Knowledge (PCK), Technological Knowledge (TK), Technological Content Knowledge (TCK), Technological Pedagogical Knowledge (TPK), and lastly, Technological Pedagogical Content Knowledge (TPCK). The TPCK framework was described as a teacher's understanding of how to coordinate the use of a subject- or topic-specific activity with a particular representation using available technologies to facilitate student learning. For instance, a science educator teaching lunar phases needs to know what the concept of lunar phases is (Content Knowledge), what exact simulation(s) is (are) available to utilize (Technological Knowledge) in order to teach this specific concept, and how this particular simulation could be integrated into the lesson plan (Pedagogical Knowledge).

How Can Teachers Develop TPCK?

One could feel that simply making teachers aware of available technological tools, well informed about the commonly used pedagogical approaches, and proficient on the content of their subject matter could be adequate for their professional development. And one could also assume that teachers, by simply combining those elements, will become successful educators in teaching with technology. However, one should also be aware of the fact that knowledge is context bound (Lave & Wenger, 1991; Lemke, 1997; Samarapungavan, Westby, & Bodner, 2006). In other words, the development of understanding is situated, and thus an understanding of TPCK can only be mastered by involving teachers in technology-, content-, and pedagogy-specific learning environments. For instance, Koehler and Mishra (2005) took a situated view and indicated in their study that encountering teachers with authentic and ill-structured problems, called Learning Technology by Design, provided an important context for their development of TPCK. By participating in design, teachers are to build a content-, technology-, and pedagogy-specific lesson plan.

In the present work, we took a comparable approach. The information provided in this chapter resulted from a project sponsored by the Scientific and Technological Research Council of Turkey. A total of 15 in-service science teachers participated in the project. One of the aims of it was to teach teachers how to integrate such tech tools as animations, simulations, video clips, models, and pictures into their instruction. The teachers participated in a 4-week training course. In the first week, they were introduced to a framework involving general guidelines for the integration of technological tools. In the present chapter, this framework will be described in detail.

Technology Integration Framework (TIF)

Technology Integration Framework (TIF) emerged when our team encountered the works of Doyle (1983, 1986, 1988) on task design. To Doyle, tasks involve four important components: (a) product(s), (b) resources, (c) operation(s), and (d) accountability. To Doyle, the term product refers to the overall goal of a task. For instance, the fossil footprint activity aims at creating an understanding that scientific knowledge is a human inference or imagination and subject to change as new evidence becomes available. The resources, on the other hand, refer to the materials to be used in the task. In the footprint activity, the material was the images of footprints (position 1, 2, and 3). Next, the operations refer to students' and teacher's roles in implementing the task. In the footprint activity, the students were to speculate on the species of animals and what is happening in each position. The teacher was, on the other hand, to show positions sequentially, ask students to reflect on each one, and provide explanations as to the linkage between students' speculations and the nature of scientific theories. Finally, the term accountability refers to the degree to which students find the task important to execute.

Being enthused by this approach, the team created a technology integration framework (TIF), which involved five major questions. The questions intended to serve teachers as general guidelines for integrating technological tools. TIF basically asked teachers to determine their topic of concern, search for available technological tools, modify the tool for their particular purpose and revise the learning objective(s) accordingly, plan students' and their own role in the integration of the tool, and finally, evaluate their classroom performance. Below, this framework is elaborated.

Technology Integration Framework (TIF):

Step 1: What Technological Tools are Available for My Topic of Concern?
    a. What are the tools offered related to my topic of concern?
    b. Which tool could serve best for my particular purpose?
Step 2: Is this Technology Appropriate for My Purpose?
  My purpose of using technology is
    a. To introduce a novel concept to students?
    b. To consolidate a particular student's conception?
    c. To stress a common misconception?

    d. To assess students' understanding of a particular subject matter?

Step 3: Do I Know Enough about It?

    a. Do I know how to handle this tool?

    b. Could my students easily operate it?

    c. What are the advantages of it?

    d. What are the limitations of it?

    e. How am I going to operate it?

Step 4: How am I Going to Integrate the Tool into My Instruction?

    a. Am I going to use the tool as replacement, amplification, or transformation?

    b. Do I know what to do in case of encountering technical problems?

    c. What is my lesson plan? Do I need to form small groups of students? Do I have a plan for supervising students?

Step 5: Has this Tool been Useful?

    a. Has this simulation been appropriate for my purpose?

    b. Have the learning goals been achieved by the students?

    c. Can this simulation be used for different purposes?

    d. Have I encountered unexpected technical problems? If yes, what precautions should I take in the future?

    e. Have I encountered pedagogical problems? If yes, what precautions should I take in the future?

    f. Do I need to make changes in my lesson plan, learning goals, or students' roles?

    g. Has this simulation led the students to get misconceptions? If yes, what precautions should I take in the future?

At this point, one could ask how TIF could be used in the integration of a particular technological tool. Then, let us follow an elementary teacher, Mrs. Ayse, who is trying to integrate a technological tool into her instruction. In the boxes are the questions from TIF that Mrs. Ayse is asking herself as guiding questions. Outside of the box is what she did in preparing, applying, and evaluating her lesson plan.

Step 1: What Technological Tools Are Available for my Topic of Concern?

Searching for an appropriate simulation via Google, Mrs. Ayse encountered the following Web site: http://phet.colorado.edu/

This Web site provided many interactive simulations for free of use. Mrs. Ayse entered the word "motion" in the search box and a list of simulations appeared. Examining those simulations, she felt she finally found what she was looking for. The simulation she found was called The Moving Man. She could either download or directly run it on the Internet. She preferred to download the program because in that way she would not need an Internet connection.

Step 2: Is This Technology Appropriate for My Purpose?

She was planning to teach her fifth graders how to read and interpret motion graphs and the simulation, the Moving Man, seemed to be appropriate for her goal.

Step 3: Do I Know Enough About it?

In order to learn how to handle the simulation, there was a link directing her to the teacher's guide. The guide was fairly helpful; and then she decided to run the simulation and operate it some time. Figure 8.2, is how it looks on her monitor.

She thought her students could easily understand and be able to operate this simulation. The simulation allowed her to simultaneously observe a moving man and the graph of his motion. In addition to that, the simulation provided motion graphs when the direction of movement was negative. "This could be an additional learning goal" she thought. Accordingly, she made a revision in her purpose and added a new learning goal into her plan. "What are the limitations of it?" she asked herself. There was only a 20-meter distance for the man to move, and he disappeared when he went farther. Moreover, she thought she did not need to use the Introduction

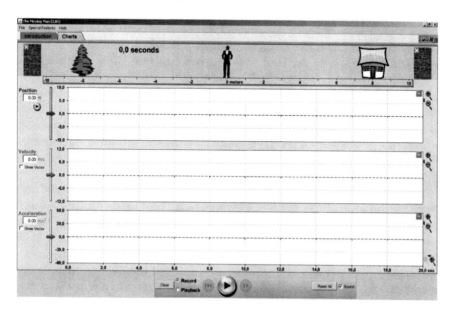

Figure 8.2   The Moving Man view (http://phet.colorado.edu/en/simulation/moving-man).

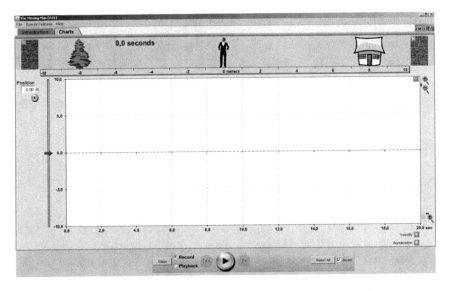

Figure 8.3 The Moving Man view without velocity and acceleration charts.

page or the velocity and acceleration charts. She thought these views could perplex her students, so she clicked on the red buttons on the right-hand side of the graphs and removed both the acceleration and velocity charts. Figure 8.3, is how it then looked.

Step 4: How Am I Going to Integrate the Tool into My Instruction?

She is going to use the tool as transformation (see Hughes, 2005 for a detailed review of this concept). It has already transformed her plans regarding her instructional method, the learning goals, and students' roles. In order to learn dealing with technical problems, she noticed a link directing her to the troubleshooting page. This page involved some information about the problems people commonly encountered running the simulations. Also, there was an e-mail address provided for communicating with a tech expert.

"Now, I am ready to prepare a lesson plan," Mrs. Ayse said silently. She noticed a link directing her to Teaching Ideas. This page involved a number of lesson plans for different student ages. Examining them, she found an appropriate lesson plan for her students and determined to use it as her lesson plan. The plan was prepared for elementary students by Sarah Borenstein. In the class, Mrs. Ayse first introduced the Moving Man to the students and showed them how to operate it. Thereafter, she asked the students to follow the directions depicted below (Figures 8.4, 8.5, and 8.6).

Moving Man Elementary Lesson

Directions:

1. Play with the *Moving Man* by dragging him back and forth. Click on the playback button at the bottom of the page to look at the graphs when you are done. Notice what is happening to the graphs as he moves. Tell your neighbor what you've noticed, and why you think it happened.

"Look, _____, I noticed that....	"Well, I think this happened because..."

   Be sure to click the "clear" button on the left side of the screen to reset your graph.

Allow students the opportunity to share with their lab partners first, then the rest of the class. Record students' ideas on a board for all students to see.

2. Look at the graph and the numbers underneath the man. Tell your neighbor what you noticed, and write or draw what you noticed.

3. Write down what the graph looks like when you drag the man towards the house. Tell your neighbor what you noticed, and why you think it happened.

GRAPH		Explanation
		Why do you think it should look like this?

4. Do you think the line will always look this way when the man is moving? Test out your ideas on the simulation. Share your thoughts with your neighbor, then write or draw what you noticed.

This part of the lesson is set up for students to make conjectures about what they see. The line will start at zero and move upwards to the right when the man is moving towards the house. Students should be allowed time to play with the simulation to see if the line will always move this way. Share and write down students' observations after they have had time to explore.

5. Write down what the graph looks like when the man is standing still. Tell your neighbor what you noticed, and why you think it happened.

GRAPH		Explanation
		Why do you think it should look like this?

Figure 8.4

6. Do you think the line will always look this way when the man is standing still? Test out your ideas on the simulation. Share your thoughts with your neighbor, then write or draw what you noticed.

   This part of the lesson is set up for students to make conjectures about what they see. The line will remain flat when the man is standing still. Some students might notice that they can tell where he is standing from the graph, as well as how long he has been standing there. Students should be allowed time to play with the simulation to see if the line will always move this way. Share and write down students' observations after they have had time to explore.

7. What do you think will happen to the line if the man moves away from the house? Draw your prediction in the space below, then test out your idea. Share what you found out with your neighbor and write it in the space marked "explanation."

GRAPH (Prediction)	Explanation
	Why do you think it should look like this?

GRAPH (Actual)	Explanation
	Why do you think it looked like this?

8. Do you think the line will always look this way when the man is moving away from the house? Test out your ideas on the simulation. Share your thoughts with your neighbor, then write or draw what you noticed.

   This part of the lesson is set up for students to make conjectures about what they see. There is also the opportunity to make a prediction based on what they noticed in the 2 previous examples. The line will move downward and to the right when the man is walking away from the house. Students should be allowed time to play with the simulation to see if the line will always move this way. Share and write down students' observations after they have had time to explore.

9. Summary
   Based on what you saw in the examples, summarize what you know.

Figure 8.5

When the graph looks like this:	Describe how the man is moving.

EXPLAIN and EXPLORE

This section allows students the opportunity to apply what they have observed in the previous section. It can be used as a formal or informal assessment on student understanding.

1. Without using *Moving Man*, draw what you think the line would look like for the following story.

   A man is napping under the tree. He wakes up and walks toward the house. He stops because he is worried that he dropped his keys. He stands still as he searches his pockets for his keys. When he discovers he can't find them, he runs towards the tree. He hits the tree and gets knocked out, so he can't move.

Graph (prediction)	Graph (actual)

2. Now use *Moving Man* to see if your graph was correct. If not, go back, and draw the correct line on the 2ⁿᵈ graph.

3. In the space below, write your own story for the following graph.

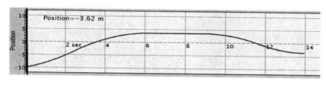

Figure 8.6

Step 5: Has this tool been useful?

In the final step of TIF, Mrs. Ayse asked herself the questions of Step 5 and took notes on the likely precautions that could be taken for her future plans.

DISCUSSION AND CONCLUSION

In conclusion, the TIF guiding questions helped Mrs. Ayse plan and create a technology-supported learning environment. In this course, she searched for an appropriate tech tool, located the one that seemed the most appropriate for the use, modified the tool according to her learning objectives, planned her own and students' roles in the integration of it, and finally evaluated her classroom performance. In this course, she not only created and used a tech tool in her classroom, she also learned the integration procedure. Therefore, the TIF guiding questions equip teachers with the knowledge of the interplay among technology, pedagogy, and content. One using TIF guiding questions recognizes the need of searching for proper technological tools related to the topic of concern, modifying the tool according to the particular learning objective, revising the learning objective(s) according to the tool when required, preplanning students' and one's own role in the integration of this particular tool, and finally, evaluating one's own classroom performance. These actions thus lead teachers to consider technology, pedagogy, and content all at once in integrating technological tools into their instruction.

REFERENCES

Akpan, J. P., & Andre, T. (2000). Using a computer simulation before dissection to help students learn anatomy. *Journal of Computers in Mathematics and Science Teaching, 19*(3), 297–313.

Barnea, N., & Dori, Y. J. (1999). High-school chemistry students' performance and gender differences in a computerized molecular modeling learning environment. *Journal of Science Education and Technology, 8*(4), 257–271.

Bell, R. L., & Trundle, K. C. (2008). The use of a computer simulation to promote scientific conceptions of moon phases. *Journal of Research in Science Teaching, 45*(3), 346–372.

Cox, S., & Graham, C. R. (2009). Diagramming TPACK in practice: Using an elaborated model of the TPACK framework to analyze and depict teacher knowledge. *TechTrends, 53*(5), 60–69.

Doyle, W. (1983). Academic work. *Review of Educational Research, 53,* 159–199.

Doyle, W. (1986). Classroom organization and management. In M. C. Witrock (Ed.), *Handbook of research on teaching* (3rd ed, pp. 392–431). New York: Macmillan.

Doyle, W. (1988). Work in mathematics classes: The context of student's thinking during instruction. *Educational Psychologist, 23,* 167–180.

Hughes, J. (2005). The role of teacher knowledge and learning experiences in forming technology-integrated pedagogy. *Journal of Technology and Teacher Education, 13*(2), 277–302.

Huppert, J., Lomask, S. M., & Lazarowitz, R. (2002). Computer simulations in the high school: Students' cognitive stages, science process skills and academic achievement in microbiology. *International Journal of Science Education, 24*(8), 803–821.

Kelly, R., & Jones, L. (2005). *A qualitative study of how general chemistry students interpret features of molecular animations.* Paper presented at the National Meeting of the American Chemical Society, Washington, DC. August 28–September 1.

Koehler, M. J., & Mishra, P. (2005). What happens when teachers design educational technology? The development of technological pedagogical content knowledge. *Journal of Educational Computing Research, 32*(2), 131–152.

Mishra, P., & Koehler, M. J. (2006). Technological pedagogical content knowledge: A framework for integrating technology in teacher knowledge. *Teachers College Record, 108*(6), 1017–1054.

Lave, J., & Wenger, E. (1991). *Situated learning: Legitimate peripheral participation.* Cambridge, UK: Cambridge University Press.

Lemke, J. L. (1997). Cognition, context, and learning: A social semiotic perspective. In D. Kirshner, & J. A. Whitson (Eds.), *Situated cognition: Social, semiotic, and psychological perspectives* (pp. 37–55). London: Lawrence Erlbaum Associates.

National Academy of Sciences. (1998). *Teaching about evolution and the nature of science.* Washington, DC: National Academy Press.

Samarapungavan, A., Westby, E., & Bodner, G. M. (2006). Contextual epistemic development in science: A comparison of chemistry students and research chemists. *Science Education, 90*(3), 468–495.

Sanger, M. J., Phelps, A. J., & Fienhold, J. (2000). Using a computer animation to improve students' conceptual understanding of a can-crushing demonstration. *Journal of Chemical Education, 77*(11), 1517–1519.

Schulman, L. S. (1986). Those who understand: Knowledge growth in teaching. *Educational Researcher, 15*(2), 414.

Schulman, L. S. (1987). Knowledge and teaching: Foundations for a new reform. *Harvard Educational Review, 57*(1), 1–22.

Wertsch, J. V. (1998). *Mind as action.* New York: Oxford University Press.

Winn, W., Stahr, F., Sarason, C., Fruland, R., Oppenheimer, P., & Lee, Y. (2006). Learning oceanography from a computer simulation compared with direct experience at sea. *Journal of Research in Science Teaching, 43*, 25–42.

CHAPTER 9

REAL-TIME EXPERIMENTS AND IMAGES (RTEI) AS OPEN LEARNING ENVIRONMENTS FOR BUILDING PHYSICS KNOWLEDGE

Giorgio Olimpo
Istituto Tecnologie Didattiche
Consiglio Nazionale Ricerche

Elena Sassi
Dipartimento Scienze Fisiche
Università di Napoli "Federico II"

ABSTRACT

Experimental laboratory activities play a key role in the process of the construction of physics knowledge, both in scientific research and in science education. Real-Time Experiments and Images (RTEI) is a type of labwork that empowers educational laboratory activities thanks to the pedagogical opportunities offered by computer acquisition of measures in real-time. The didac-

Contemporary Science Teaching Approaches, pages 179–212
Copyright © 2012 by Information Age Publishing
179

tic value of RTEI is nowadays widely acknowledged, the field having been the object of research and experimentations for many years.

The role of labwork to overcome some inadequacies of current physics education is discussed and the specific educational opportunities offered by RTEI are outlined.

These potentialities can be fully achieved only when RTEI is interpreted as an Open Learning Environment (OLE), where the learners are the main actors of the knowledge construction process. In this case, RTEI approaches allow addressing different levels of knowledge, ranging from topical knowledge (related to a specific topic) to networked knowledge (related to the links among different topics and abilities) to meta-knowledge (related to the ability to build new knowledge). Some examples illustrating the pedagogical wealth of RTEI OLE are presented in the area of Classical Mechanics.

THE LABORATORY WORK IN PHYSICS EDUCATION

Experimental activities carried out through laboratory work (from now on labwork) are a crucial component of physics education. They may significantly improve the quality of the teaching/learning processes leading to the construction of a body of knowledge that is sound, stable, and capable of fostering further expansions. Actually, labwork, if intended as an open-learning environment, may become a fundamental tool to familiarize the learners with the nature of physics as an experimental science aimed at observing, describing, and modeling our natural universe. At the same time, it may foster the interest in science and the motivation of learners[1] thanks to the possibility of manipulating specific experimental environments, investigating the behavior of real phenomena, and putting abstract ideas to work. Besides, labwork can contribute to clarifying the image of science and improving the scientific communication skills of learners.

To make further explicit the meaning of labwork, it may be helpful to explore the analogy between two quite distinct fields: research in physics and physics education. According to this analogy (see Figure 9.1), in both areas, there are communities involved in the construction of physics knowledge. In the case of research, the community of physicists is active in expanding the boundaries of existing knowledge, while in the case of education, the community (for instance, a school class with students and teachers) aims at expanding the boundaries of learners' knowledge. The knowledge that is built "in the brain" of learners, strictly speaking, is not a new knowledge, since it is already part of the body of knowledge developed by the community of physicists. But it is important to highlight that at the very core of both situations there is a process of construction of physical knowledge. Based on this similarity, it is reasonable to assume that both situations share,

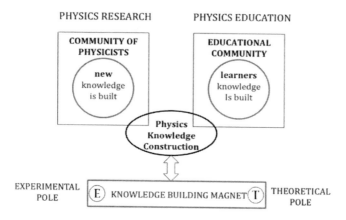

Figure 9.1   The analogy between physics research and physics education.

at least in principle, a common method for knowledge construction. Otherwise, the physics that is learned at school would be nothing more than a collection of laws governing the phenomena without the comprehension of the processes needed to generate new knowledge and of the general structure of physics as a discipline, which corresponds to a very reductive vision of physics education. The method for constructing physics knowledge is the classical Galilean method, whose main feature is the existence of two complementary synergistically interacting polarities: the theoretical and the experimental one. It is true that some physicists work mostly on the theoretical and others on the experimental side, but any new knowledge is the result of the work of a community where both aspects are present. In Figure 9.1, the method for building physical knowledge is represented as a sort of magnet, whose poles are marked with experimental and theoretical attributes. Like in a real magnet, these two poles are inseparable: theoretical studies almost always have to refer to phenomena or experimental results, and a purely experimental activity without theoretical reflections can hardly make sense. This *knowledge building magnet* ought to also be at the very heart of any process aimed at physics education. Exclusive use of purely narrative approaches in teaching sounds like an attempt to separate two poles of a magnet and is one of the major reasons for the poor performance of many school systems in the area of physics education, as shown by international studies. Some projects have proposed significant integration of labwork in their structure; for instance, the PSSC[2] developed in the United States (e.g., French, 1986) and the Nuffield Physics Project, developed in the UK (for an initial description, see Maddox, 1966). Both these projects have been successfully used and/or adapted in many countries.

An approach to physics education based on labwork is often qualified with the attribute "hands-on," used to mark a difference and in distance from narrative approaches and to make the educational value of practical operations in experimental environments more explicit. However, in accordance with the above considerations, the desirable approach to physics education is better qualified as "hands-on and minds-on," because these two polarities shape and frame both physics research and physics education, so the synergy between hands-on and minds-on is a key factor in both fields. Besides, the trap of naïve empiricism is always present, and a well-weighted combination of hands-on activities and minds-on reflections is of paramount importance to avoid it (Sassi, 2005).

It is worth mentioning that adding some labwork to traditional teaching may not be enough. Some naïve interpretations and/or misunderstanding about the role and the meaning of labwork in physics (and science) education are robust and still common among teachers and students. Often, experimental activities are interpreted in a rather reductive way as ancillary educational activities, less important than lectures and presentation, mainly aimed at acquiring manual skills or collecting and analyzing some data (very often from ready-to-go traditional apparatuses). The method is that of providing detailed instructions and, in some cases, step-by-step guidance either directly through the teacher or through labworksheets. About 10 years ago, the EU Project, Labwork in Science Education (Séré et al., 1998), clearly showed that this attitude toward labwork was quite common in several EU countries: the labwork was generally recognized as an important didactic moment, but still the experimental and the theoretical poles did not assume the same level of dignity in the learning construction process. In past years, many papers have been published about the role and meaning of labwork and its contribution to physics education (e.g., Hofstein & Lunetta, 1982, 2004; Trumper 2003). In 2008, the International Commission on Physics Education (ICPE) published online a second volume about "Connecting Research in Physics Education with Teacher Education," (http://web.phys.ksu.edu/icpe/Publications). Some papers specifically discuss labwork (Sassi & Vicentini, 2008; Thornton, 2008).

OVERCOMING OBSTACLES AND INADEQUACIES IN PHYSICS EDUCATION: THE ROLE OF LABWORK

A deeply rooted teaching tradition tends to present physics mainly as a collection of laws and related formulas, while experimental activities are not much valued and practiced. This tradition is supported by a variety of factors, but two of them seem to be particularly influential: (a) dealing with experimental activities, where unforeseen events may easily happen, on the average, more demanding for the teacher than presenting a topic in a traditional way; and (b) that many physics teachers have not had significant labwork experience in their preservice education (e.g., in Italy, many in-service teachers with a university degree in mathematics are not familiar with physics labwork).

The above-mentioned approach is among the causes of the students' inadequate preparation, as shown by international surveys going on for many years, like PISA (OECD Program for International Student Assessment, http://www.pisa.oecd.org) and TIMSS (Trends in International Mathematics and Science Study, http://rnces.ed.gov/timss). This approach fails to foster not only students' understanding but also interest and motivation, and reinforces the idea that physics (and more generally, science) is a subject reserved for a minority of gifted people. On the contrary, physics education research has shown that a didactic methodology based on a wise integration of labwork with theoretical reflection can give substantial contributions to overcome many of the problems and inadequacies that characterize the present practice of physics education. In the following, the main opportunities provided by this integration are briefly discussed.

FROM NARRATIVE TO PARTICIPATIVE APPROACHES

It is quite common to present physics as a sort of narration dealing with phenomena, laws, formulas, and anecdotes about physicists of the past, etc. In the worst case, the narrative approach does not use a rigorous language and presents uncorrelated concepts, not favoring and supporting the development of critical thinking. Even when some labwork is included, the transmissive "chalk and talk" style is predominant, often when some technological support is used.

This is a sort of distortion of the very nature of physics which, as we have already mentioned, involves a very articulated interplay of experiments, models, laws, and theories that lead to the construction of physical knowledge. Even assessment, especially in basic physics courses, is very often based either on narration or on the solution of standard problems requiring a mere application of formulas. This encourages mainly the rote memorization of fragmented notions. Narration is a facet of the transmissive teaching, an approach based on the belief that teacher transmission of knowledge is sufficient and effective. This epistemology conflicts with the constructivist model of learning (Rieber & Carton, 1987), claiming that learners actively build their knowledge, helped by interacting with peers, by direct experiences and by having time and occasions to reflect about the learning goals. The role of teachers therefore changes from authoritative transmitters to helpers and mediators of knowledge. Educational research has shown that by shaping the learning process as a rich and meaningful interplay between labwork and theoretical reflection and assigning to the learner the role of actor of the process of construction of physical knowledge, is it possible to foster the construction of a personal knowledge net, to create motivation, and to make physics education more effective and useful. Just by listening to narrations, students acquire only a little understanding and easily forget, while by preparing experiments, analyzing and modeling measures, discussing hypotheses and links among phenomena, becoming aware of framing theories, etc., they become concretely aware of goals, difficulties, practical and conceptual problems of physics, and are strongly facilitated in interiorizing both processes and results. A typical example of how labwork can influence the learning process is that of the laws of ideal cases. So often they are taken as a starting point of the teaching process. This approach, which is fed by several textbooks, hides the complex and multifaceted path followed to establish ideal laws and prevents the student from having a personal experience of the *path* from "the real to the ideal," whose value is very high from both the disciplinary and cognitive standpoints.

Labwork favors a participative approach not only at an individual level but also at a group level. This reinforces relational and collaborative attitudes, fosters peer learning (sometimes the students are very familiar with technology-based tools), and is of a high cognitive value since the group is a source of different ideas (preconceptions, viewpoints, hypotheses) and abilities, which, under the wise guidance of the teacher, may become actual resources for the learning process. Of course the teachers must abandon the role of narrator and become for the group a sort of point of reference who suggests, orients, asks key questions, points out contradictions, and supports the group in a variety of ways. Narration is not always negative, sometimes it can provide a framework, entertain, and create interest and motivation. What is to be avoided is the use of narration as the main instrument of teaching.

Finally, it should be mentioned that the increase of student motivation, which is one of the benefits of a labwork-based participative approach in physics education, may contribute to fight the decrease of interest in scientific subjects and related jobs/careers, which is a growing problem in industrialized countries. The ROSE project (Sjøberg & Schreiner, 2010) has shown that school science for students up to 15 years (in 40 countries) is "less interesting than other subjects,... has not opened my eyes for new and exciting jobs,... has not increased my career chances,... has not increased my appreciation for nature,... has not shown me the importance of S&T for our way of living."

FROM COMMONSENSE KNOWLEDGE
TO PHYSICS KNOWLEDGE

Some common learning obstacles encountered by learners when studying physics are related to naïve ideas and reasoning patterns coming from commonsense knowledge and conflicting with disciplinary knowledge. As a very simple example, we may quote the difficulties created by the different meaning that some terms such as force, energy, temperature, heat, field, ray, wave, etc. have in everyday language with respect to their definitions in physics.

However, commonsense knowledge should not be considered as something negative and undesirable. In fact, it directly derives from the daily interaction with our natural environment and is appropriate and even necessary for everyday life. What is critical in physics education is that the nature of commonsense knowledge, which is the initial patrimony of learners, is quite different from scientific knowledge: it is context-dependent, focused on practical values, based on skills, incomplete, mostly qualitative, accepting of contradictions, rooted in personal experience and therefore resistant to change, and generally expressed in natural language. On the contrary, scientific knowledge is developed by the scientific community through an intentional process, is focused on speculative values, is self-consistent (contradictions are not accepted), aims at completeness, and is expressed in formal mathematical languages. It is only natural for learners trying to transfer conceptions and mental models almost unconsciously derived from their experience to the area of scientific knowledge. In fact, the learning difficulties arising from the clash between these two types of knowledge show deep similarities in different countries, types of school, and cultural contexts (for a bibliography, see Duit, 2009).

We have seen that the free interaction with the natural environment is one of the sources of commonsense knowledge; in a similar way, a gently "assisted" interaction with the dedicated environments provided by the laboratory can be a very important base from which to foster scientific knowledge (see Figure 9.2). The figure shows that the labwork may assume

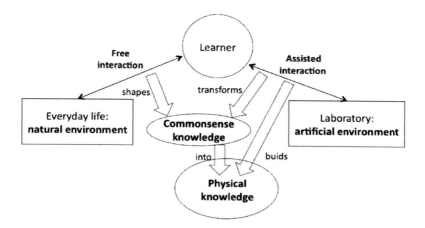

Figure 9.2 The laboratory as a dedicated environment to build physical knowledge.

two different, but not necessarily separate, meanings: building scientific knowledge and transforming the commonsense conceptions (or reasoning patterns) conflicting with physics into scientific knowledge. In fact, the study of physics phenomena in a specific and dedicated experimental space allows the learners to become familiar with fundamental aspects of physics knowledge: how to explore a phenomenology, how to collect and analyze data, how to model data, how to interpret models, how to integrate the output of labwork with previous knowledge, etc. Eliciting learners' ideas about the phenomena under study is an important—very often should be the first—moment in any lab activity since it allows the possible identification of naïve ideas, addressing them mainly by working through experiments and transforming them into disciplinary knowledge. Of course, the teacher has a key role in helping learners to make their ideas/reasoning explicit, to educate their critical-thinking skills, and to orient the labwork from both the conceptual and practical standpoints.

FROM SEALED COMPARTMENTS
TO NETWORKED KNOWLEDGE

One of the typical drawbacks of the narrative approach discussed above is that the knowledge proposed to the learners is a collection of topics without concrete and meaningful links among them. Often, when these links are proposed, they have little impact on learners since they are just another part of the narration. We may refer to this approach as "sealed compartments" acquisition of notions. This trend may depend on many reasons, but two of them seem to be more relevant than others: on one side, teachers tend to perpetuate the cultural "tradition" they have been familiar with since the time they were students; on the other, they may be unwilling to enter new territories that require extra work (redesigning the course, collaborating with other teachers, studying new subjects, etc.) and a drastic change of self-image (from teachers who know all the answers and know how to deal with any situation to a more expert member of a group of learners who may need to investigate with the rest of the group to answer questions and find solutions). In fact, establishing links among different areas of knowledge often requires traveling unfamiliar territories. This typically happens when ICT-based experimental setups are used. In these cases, learners may be more familiar with technology manipulation than teachers, and a new attitude of reliance on students' skills is required of teachers.

Labwork may significantly contribute to the process of establishing meaningful links among different topics. For instance, experiments about conservation of energy in multiple bounces of a rubber ball on the floor can help to connect the concept of ideal elastic collision with the unavoidable "dirty" effects due to air resistance, energy dissipation within the ball material, sound emission due to the impact on the floor, and the consequent decrease in the maximum height reached by the ball that is visible to the eye. The technical aspects of experiments may help to establish further connections among different topics or disciplines. For instance, a reflection about how the same technique (e.g., measurements of electric potential difference) is used for studying many physics or chemistry phenomena can help the construction of networked knowledge. The skills involved in using a specific instrument can also contribute to this networking. For instance, a calorimeter is used for measurements in several content areas; so to discuss the basics of how it works, its potentialities and limits, how to interpret calorimetric results, etc. is a way to connect thermal phenomena in physics to chemical and biological ones. Finally, mathematical procedures can also offer occasions to build links, for example, data-fitting is widely common in labwork together with some numerical methods used in some modeling activities.

FROM RITUALS TO EFFECTIVE TEACHING AND LEARNING PROCESSES

Some deep-rooted, undesirable teaching habits should also be taken into account (Viennot, 2003a, 2003b, 2006). Some of these habits may be related to naïve epistemologies, often not even conscious ones (e.g., it is appropriate/sufficient to repeat the textbook); to convictions/beliefs (e.g., I teach as I have been taught); to blaming habits (e.g., students do not understand because of their lack of intelligence and commitment; the previous school level has not adequately prepared the students); to inaccurate didactic materials that perpetuate ineffective presentations (sometimes also with mistakes); to "gray zones" in the subject-matter knowledge corresponding to contents unfamiliar or not fully understood; to very little or no familiarity with results/proposals coming from Physics Education Research (PER), and so on. Also, some rituals are present on the learners' side; in part linked to the previous teaching rituals, in part due to the widening gap between school and society. For instance, learners practice both rote memorization of fragmented notions (just to pass the task or the exam) and a fast forgetting of the swallowed notions.

The use of ideal case laws as starting points is another very common practice. This is a serious obstacle since most learners have not developed yet the abstraction abilities required to link these cases with everyday-life experiences; furthermore, starting with ideal cases may easily obscure the fact that in the development of physics, these laws have been, almost always, the endpoint of complex and much-articulated historical paths, influenced by the cultural and socioeconomic features of the context.

Labwork may contribute to address the above mentioned teachers' and learners' rituals. For the latter, an increase in interest and motivation may easily derive from their engagement in all phases of the experiments, especially from deciding which questions to investigate; from building (at least partially) the setup; in modeling the collected data; from discussing advantages and limits of the models used; i.e., on all those phases usually not much practiced in traditional interpretations of laboratory activities. On the teacher's side, an intentional effort in organizing the labwork as an open-learning environment can foster important changes both in his/her disciplinary expertise and in conceptions about teaching/learning processes. While coping with unforeseen events, which commonly happen in an *open* laboratory, a reconstruction of teacher knowledge may occur. When the teacher is available to accept learners' expertise, especially on aspects where s/he may be less competent, learning becomes more similar to a collaborative enterprise with a positive impact on motivation. While experiencing learners' naïve ideas or reasoning patterns conflicting with disciplinary knowledge, the teacher may become more aware of the need to become familiar with results and proposals provided by PER.

THE POTENTIAL OF RTEI APPROACHES

Here we discuss those experimental activities dealing with real phenomena and based on the use of probes/sensors directly connected to a computer through an interface that converts analogical signals into digital information. The acronym RTEI stands for Real Time Experiments and Images; this last word is explicitly inserted into the acronym to point out the role of iconic representations of data (usually time graphs), one of the key features of this type of labwork.

Other types of technology-based laboratories such as Remotely Controlled Laboratories (where experiments on setups located somewhere in the world can be performed via the Internet) or Virtual Laboratories (where virtual apparatuses can be manipulated to obtain data) will not be discussed.

RTEI systems have been around since the early 80s and are based on mature technologies; several systems are available, for example, Vernier Software USA (www. vernier.com); Pasco Scientific USA (www.pasco.com); COACH http://www.cma.science.uva.nl/english/index.html). The main advantages of this approach have been recognized since the beginning: learners' contact with real phenomena, immediate feedback via measurement graphs, increased motivation thanks to teamwork, powerful data collection, and representations facilities that speed up the whole experimental process (Thornton, 1987; Thornton & Sokoloff, 1998).

RTEI is quite different from the traditional "ready-to-go" experimental apparatuses commonly used in many schools. In most cases, these apparatuses are focused on specific experiments or families of experiments and on the collection/analysis of predefined sets of data. Actually, RTEI make provision for open, student-centered learning environments, which can be freely configured and adapted to different phenomena, experimental situations, and learning objectives. This flexibility allows not only adapting to many pedagogical plans but also coping with different students' needs and cognitive styles and/or with unforeseen situations emerging during the enactment of a plan.

The Main Technical Features of RTEI

The above flexibility is eased and fostered by the following RTEI technical features:

- A variety of sensors for different physical quantities (e.g., position, force, temperature, electric current and voltage, pressure, etc.) allow the exploration of a wide variety of phenomena.
- Large amounts of measurements can be easily collected (rate is usually some tens/second). This feature offers a double advantage: on one side, it drastically reduces the time needed for measurements and, on the other, allows the exploration of transients and, more generally, very fast phenomena
- Setups, conditions, variables, and parameters of an experiment can be easily changed. This eases the adoption of variational approaches ("what happens if . . . varies?), which are of a high pedagogical value
- Multi-representations of data (one of the key functionalities of RTEI) can be exploited when collecting data and analyzing collected data. Before measuring, it is possible to choose the types and numbers of graphs to display in real time (i.e., while the phenomenon in study is happening). This allows the exploration of the time evolution of phenomena, which is particularly relevant for the study of transients very fast and very slow processes. Multi-representations of the collected measures is also very useful; data can be easily and rapidly represented in a variety of different ways as, for example, variables on graphs' axes, numerical tables, special location of graphs for co-relational purposes, comparison of data from different experiments, etc.). Therefore, learners' abilities to manipulate data, to reason on data representations, and to choose the representaional scheme most appropriate for the aspect to analyze can be easily fostered and enhanced.
- Data fitting and modeling facilities are provided. These features are very useful: the young learners may not have the mathemati-

cal knowledge needed for the data fitting and to build a modeling environment is not straightforward; moreover, when the experimental data are very many, the computations involved may be practically impossible to perform manually. Dedicated environments are therefore needed. Sometimes the above features may also be found in non RTEI apparatuses

These technical features make provision for different complementary educational opportunities: to explore a specific phenomenon and its features; to observe which changes in setup, variables, and parameters of the experiment are significant and what are their effects; to obtain quantitative information and to infer regularities to summarize in rules and models; to check an already-made measure; to investigate a model or a law already studied; etc.

Typically, an RTEI activity starts by identifying a phenomenon to explore and the related questions to be answered; goes on with the building of the setup; proceeds by exploring the phenomenology and inferring regularities and rules, collecting data, and observing their representations; and ends with data modeling and discussing possible ideal case laws (if it is the case). In all the phases, the learner should play an active and constructive role, even though not always autonomous. This is a key condition to achieve a high degree of educational effectiveness. Actually, the RTEI systems can also be used in a very directive way according to step-by-step instructions, but this approach does not take full advantage of the educational potential of RTEI systems, which are intrinsically Open Learning Environments (OLE).

THE EDUCATIONAL POTENTIAL OF RTEI

RTEI as OLE may trigger different educational dynamics, which in turn, may influence the learners at different cognitive levels. In this section, we briefly discuss those dynamics and identify three level of knowledge which may be fostered and enhanced by RTEI approaches.

The PEC Learning Cycle and its Educational Facets

One of the most important features of RTEI OLE is to facilitate and support the practice of the PEC (Prediction-Experiment-Comparison) cycle, thanks to the technical features discussed above that are peculiar to RTEI and not present in traditional labwork. This well-known learning cycle, together with similar ones,[3] allows the learners to familiarize with important aspects of physics methodology and to manipulate both objects and ideas

(Champagne, Gunstone, & Klopfer, 1985; Chinn & Brewer, 1993; Linn & Eylon, 2006; White & Gunstone, 1992). The Prediction phase elicits the learners' ideas/reasoning and the difficulties to address. The Experiment phase deals with exploration of phenomenology; possible conflicts between the predictions and what is measurable; measures and their analysis; distinction of relevant aspects from other effects; identification of regularities/rules; and data multi-representations, fitting, and modeling. The Comparison phase analyzes and compares the previous results. The cycle can be repeated, for example, to address discrepancies between Prediction and Experiments and to ease the convergence toward stable learning objectives.

Working with the PEC cycle by RTEI OLE facilitates the adoption of specific approaches of a high innovative value for the physics education practice. Three of them are briefly discussed.

1. Variational approach: "What happens if... changes ?"

An important aspect of labwork relevant for the construction of critical thinking and formal knowledge is to understand which variables, parameters, and conditions do influence (or not) the development of the phenomenon of study. RTEI OLE approaches allow/facilitate a variational analysis since fast repetitions of experiments in different conditions are practicable. The influence of specific factors on experiment's results can be identified with a "What happens if... changes?" approach which is characterized by a high cognitive and motivational value. The learners are also helped in discriminating among different co-present effects and in deciding, according to the objectives aimed at, which "secondary" ones to minimize. For instance, when studying the heating of a liquid by a constant power source and the aimed-at ideal case is a linear increase of its temperature, the heat transport through the container's lateral surface and the evaporation at the free surface of the liquid can be considered as "secondary" effects to minimize. A variational approach can also be practiced when using simulations; here, what can happen is only what has been planned by the software author. Alternately, in an experiment, unforeseen events may occur, for instance, related to the hardware/software of the apparatus; possible misunderstandings about the contents addressed; right or wrong evocations of other phenomena, etc. All these events can be used as resources for clarifying/deepening the topic addressed or linking with other topics/areas or fixing possible setup problems.

2. The "Real to Ideal" approach: from complex familiar phenomena to ideal cases/laws

The paths start from studying complex phenomena, familiar to learners and known in terms of commonsense knowledge and move toward abstract ideal cases and their laws. Starting with experiments about everyday complex phenomena allows: eliciting the learners' naïve physics ideas; using their perceptive knowledge as a resource; and avoiding the immediate impact with ideal physics case laws that are far from the common experience of learners, require not trivial capabilities of abstraction and are perceived (especially by young students) as difficult or strange. The experimental study of these familiar phenomena aims at identifying/confirming phenomenological regularities, codifying them in rules (in mathematical language, as much as possible), and modeling them with quantitative models. This process implies the minimizing of "secondary" effects with respect to the ideal cases aimed at, for example, the effects of friction when the paths start studying a person's walk and aims at the ideal case of uniform motion of a point-like object on a no-friction track. The steps in proceeding toward "cleaner, simplified" motions, with respect to regular walks, can be: walking by sliding the feet (to reduce intrinsic discontinuities of walking by steps), and a wheeled cart on a floor, then on a low friction track. The modeling starts with data-fitting and may proceed with activities in modeling environments. The final step is the abstraction of the ideal case/model with its formal law.

3. The Knowledge Integration approach

A deep and sound physical (and more generally scientific) knowledge requires the integraion of different types of knowledge that may belong to different levels of the human consciousness and action. This integration originates the fullness of understanding and the ability to make provisions and to act intentionally on the phenomenological reality. However, this integration is not much valued and practiced in the current teaching practice of basic physics. RTEI OLE offers many ways to integrate several types of knowledge as: perceptual knowledge through experiments with strong links to perception (particularly effective in the study of motion-force and thermal phenomena); commonsense knowledge by eliciting naïve physics ideas/reasoning rapidly comparable with experiments' results; experimental knowledge via building/modifying experiments' setups; representational knowledge by working with multi-representations of the same data and choosing the most suitable representation for the problem to solve; variational knowledge by analyzing the effects of changes in conditions/ parameters of the experiment; and modeling knowledge by data fitting, use of modeling environments, interpretations of models, their description power and limits.

This knowledge integration is a source of educational flexibility helping both peer learning and transforming the teacher/learners group into a community of practice aiming at a common objective.

Three Different Cognitive Dimensions

The RTEI OLE approaches, when implemented according to their full potentialities, allow addressing three cognitive dimensions (or levels) corresponding to the construction of different types of knowledge, which can be referred to as *topical knowledge, networked knowledge,* and *meta-knowledge.* These levels, though conceptually distinct, in practice cannot be dealt with separately because they are present, in different measure, in any RTEI OLE teaching/learning activity.

Topical knowledge. This level of knowledge is focused on specific physical topics: crucial concepts about a phenomenon, its regularities and its aspects possibly conflicting with commonsense knowledge, its interpreting model(s), its governing laws and their application to specific cases, etc. This level of knowledge may also include the design and the construction of experimental setups for studying that phenomenon. RTEI OLE approaches can significantly improve the quality of topical knowledge because they imply a constructive path that builds and interconnects all the different elements of this type of knowledge. Learners are involved in all phases of a cycle that usually starts with the elicitation of their preexistent knowledge and continues with design and construction of the experimental setup, with experimental data collection, and with the analysis and interpretations of those data. The final, very important step is the comparison between the experimental results and the knowledge previously elicited (also in the predictions about the experiments), which may confirm the learner's previous knowledge or conflict with it. Of course, several iterations of the PEC cycle may be required for a gradual achievement of the correct physical knowledge. This process gives meaning to each individual step and contributes to building an integrated knowledge, where the links among different elements of that knowledge have been personally experienced by the learners. Possible type of questions fostering this level of knowledge may be: How do you expect...will behave? Why do you think...about this phenomenon? How do you justify this behavior? How would you check...? How would you modify the experiment to...?

Networked knowledge. This second level refers to the construction of a knowledge network where different types of concepts and skills are linked together. This network may involve a variety of topics belonging to different areas such as physics (and more generally, science), technology, and applications in different fields, such as history, etc. Figure 9.3 shows some sources of links that are very common and natural opportunities in RTEI OLE approaches.

Connections among the topic of study and other physics (and more generally, scientific) topics may be fostered by a variety of factors, such as co-

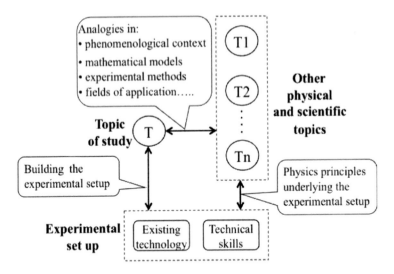

Figure 9.3 Typical sources of connections in RTEI OLE approaches.

presence within the same phenomenology or the same application field, shared mathematical models, similar experimental approaches, etc.

In the process of building the experimental setup for the topic of study, connections with technological topics referring to the equipment being used are easily established, and the related technical and operational skills are naturally trained.

Finally, the experimental setup has direct connections with physics topics different from the one being studied. For instance, position sensors based on sonar provide the opportunity for a direct connection with ultrasonic waves and their behavior; Hall effect force sensors allow the establishment of links with magnetic fields; temperature sensors based on thermocouples may point to electric properties of materials; and so on. Possible questions to stimulate the construction of networked knowledge can be: Which forces are active in this phenomenon? Where did you encounter a similar behavior? Which analogy (similarities plus differences) can you establish between this phenomenon and . . . ? Which physical phenomenon could be used for measuring . . . ?

Meta-knowledge. Here the term meta-knowledge is mainly intended as the ability to build new knowledge in the area of physics, both theoretical and experimental; of course, from the viewpoint of a learner in the framework of a learning process and not of a scientist exploring the unknown. Meta-knowledge is a condition for building both topical and networked knowledge and, at the same time, the process of knowledge building, carried on within a teacher-assisted environment, fosters the growth of learners' meta-knowledge.

Meta-knowledge does not consist simply of the application of a specific predetermined method or set of methods. In fact, it is largely heuristic in nature and strongly depends on the extension of the knowledge (both topical and networked) the learner has already built and on the richness and the quality of the experiences of knowledge building s/he has gone through.

In general, meta-knowledge cannot be represented in a procedural way; however it is possible to identify the main abilities that are necessary components of this type of knowledge. Some important items on the list are asking meaningful questions, recognizing analogies, identifying conflicts, distinguishing accessory from fundamental effects (as Galileo claimed: "neglect all that is considered contingent and accessory in order to be able to generalize and quantify"), adapting previously used methods, interpreting data, etc. These abilities, being part of meta-knowledge, are necessary conceptual tools for building topical and networked knowledge.

Figure 9.4 shows that meta-knowledge is not learned as an abstract subject but can only grow contextually with specific topical and networked knowledge.

Meta-knowledge implies the skill to apply abstract abilities to specific, concrete physical cases; usually a young learner has little or no familiarity with this skill. It is the teacher's responsibility to create favorable situations and to prompt the learners in such a way that the meta-knowledge required for the achievement of specific topics and connections may emerge. It is also the teacher's responsibility to stimulate reflection on the path that has been followed so as to foster understanding of the conceptual, abstract

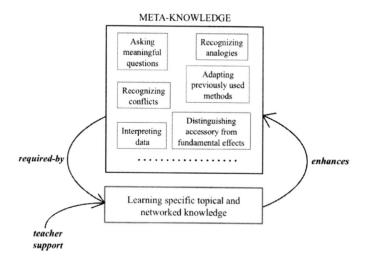

Figure 9.4 The relation of meta-knowledge with topical and networked knowledge.

meaning of specific questions, choices, and solutions in order to enhance learners' meta-knowledge.

Possible stimuli related to this level of knowledge can be: Explain why this result conflicts with your initial hypotheses. Which ideas have you changed after having studied...and why? Which physical effect can originate this data trend? Which effect should you minimize to approximate the ideal behavior of...? Describe in your own words the method we have followed.

RTEI OLE: SOME EMBLEMATIC EXAMPLES

These examples address topics taught in almost all Classical Mechanics courses. Many related learning obstacles have been researched, and the RTEI OLE approaches allow addressing them. Here it is assumed that the learners are familiar with the system used; if not, the motion and force sensors have to be introduced. The comments are hints for a script that teacher and learners will act together according to the objectives aimed at and after a didactical adaptation to the specific context. These activities have been experimented with students (15–18 years) and in both preservice and in-service teachers' education programs (Pinto et al. 2003; Sassi, 2002; Testa, Monroy, & Sassi, 2002). Recently, PHYSWARE: A Collaborative Workshop on Low-cost Equipment and Appropriate Technologies that Promote Undergraduate Level, Hands-on Physics Education throughout the Developing World (ICTP, Trieste, Italy 2009), (http://cdsagenda5.ictp.trieste.it/full_display.php?smr=0&ida=a07137) has involved about 40 physics teacher-educators from about 30 nations, for 2 weeks, in full immersion modality. From 2012 on, PHYSWARE will continue in developing countries and at ICPT.

Cart Up/Down a Ramp: Acceleration, Friction, Impulsive Forces

This example is discussed in detail to clarify its rationale. The content is usually taught as a key example of constant acceleration motion; many textbooks and teachers start from the ideal case of a no-friction ramp-car system and use only diagrams and formulas. This approach requires a familiarity with the abstraction process needed to extrapolate from a familiar, complex phenomenon (where friction plays an important role) to an ideal no-friction case. In common experience, friction is experienced continuously; we *know* that without friction it is impossible to walk, even if this knowledge does not imply the ability to express formally the friction force and its effects. Similarly, the damping effects are *known* independently from reasoning about the complex concept of energy, its conservation law, and what dissipation is. Examples of learning problems commonly encountered in studying how a cart moves on a ramp are: acceleration is present only when a cart moves down; when a cart at the ramp bottom starts moving up, it is because of a "very fast" force; a cart inverts its motion when its "force capital" is finished, at inversion no force acts; friction is "part" of motion; etc.

The proposed conceptual path starts with the Prediction phase of the PEC cycle to elicit the learners' ideas/reasoning. Example questions that may help are: Describe how and why a cart moves up/down a ramp? What makes a cart at the ramp bottom to start moving up? What happens when the cart invert its motion? Why does the cart stop after a while? Which results do you expect by the experiment? etc. The predictions are analyzed and kept. Then the Experiment phase starts by discussing which aspects to focus on, the questions the experiment(s) can answer, the set up and the position of the ultrasonic sensor (at top or bottom of the ramp), etc. Having built the setup, a first experiment can conveniently measure the motion of a cart starting from the ramp bottom having been given an initial velocity by a fast kick. If needed, it may be useful to clarify the conventions used in Cartesian coordinate systems: axes directions, origin position, etc. Here the origin is in the sensor, and the positive direction of the x axis corresponds to the increasing distance from the sensor. Common difficulties about the sign of velocity and acceleration can also be addressed. The velocity values are calculated via a finite difference method; this allows the introduction of its basic elements and the clarification of the resulting enhancement of possible variations in the primary data (position and time measurements).

Typical kinematical graphs of the cart up/down a ramp are shown in Figure 9.5.

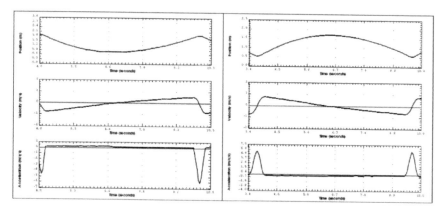

Figure 9.5 An up/down motion of a cart on a ramp. Left (right) sensor at ramp top (bottom). The effects of the change in the origin of the coordinate system are evident.

The nonlinear trend shape of position versus time s(t) is evident, together with a (quasi) linear v(t) and (quasi) constant a(t). It is useful to correlate the v(t) graphs with s(t) and a(t); this makes it easy to understand that a motion inversion is marked by a zero velocity value and that the short duration of the a(t) peaks indicates the fast kick giving the cart its initial velocity. Questions about the diverse slopes in v(t), the peaks and values in a(t), appropriate data fit, etc. can help the learning dynamics.

According to the learners' mathematical knowledge, the (quasi) linear trend of v(t) can be correlated with the (quasi) parabolic trend of s(t) and the (quasi) constant value of a(t). Also, paper/pencil tasks on printouts of RTEI graphs are useful. As for any image, it is necessary to know the iconic code in order to interpret it. A graph with its caption conveys most of the important features of sensor-based experiments and helps the learners become familiar with still images, different and complementary with respect to the graphs of ideal cases or of mathematical functions, common in textbooks.

A possible conceptual and practical path, linked to the above *topical knowledge* is: learners prepare the setup and check that the still cart is *seen* by the sensor; sensor off: the cart up/down motion is observed; predictions (words, sketches) about s(t), v(t) and a(t) are expressed, with ideas about friction and its role; sensor on: experiments are performed; representations of measures are optimized and compared with predictions, agreements/discrepancies are discussed; values and signs of acceleration are discussed, with special attention to ideas about negative (positive) acceleration in the up (down) motion; estimation of average acceleration by v(t) slopes; correlation between v(t) and a(t) focusing on slope changes in v(t) and values' variations in a(t); ideas/experiences about friction are discussed, focusing on friction as a force "opposing the motion"; experiments with friction increased; convergence on the friction force direction opposite to velocity and here proportional to it; and in the ideal case of no-friction, the acceleration value is the same fraction of g in up/down motion.

This path can expand to the friction's effects on s(t), v(t) and a(t), starting with a s(t) fit (Figure 9.6).

For better readability, the fit of the complete up/down motion is not shown, it does not represent well the up or the down motion. The parabola of the left (right) fit describes well only the up (down) motion. Switching between the formal language of mathematics and the phenomenological language of the measures allows linking areas of knowledge and making "visible" features not immediately quantifiable. The parabolic function is symmetrical with respect to its axis; it shrinks (opens) when the coefficient of the square term (here representing acceleration) increases (decreases). This approach links two knowledge areas and help the learners become aware of the physics and mathematics interplay.

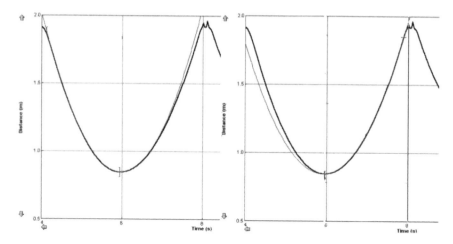

Figure 9.6   Two copies of s(t) (sensor at ramp top; 20 measures/s). The left (right) side represents the up (down) motion of the cart. Dark curves are parabolic fits. In the left (right) graph, only the up (down) motion data have been fitted.

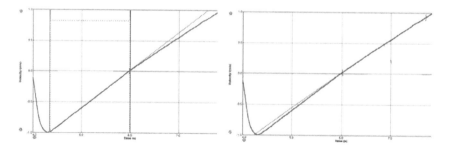

Figure 9.7   Two copies of v(t); light lines are linear fits.

Figure 9.7 shows the analysis of up/down motion in terms of v(t) (see also Larson, 1999). The velocity values are positive (negative) in the down (up) motion; zero in v(t) represents the cart motion inversion at its maximum height on the ramp; the slopes' variations are clearly seen. A linear fit represents well the up (down) motion, but not both. The steeper fit at left indicates greater acceleration in the up motion, as confirmed by the slope of the fit at right. In a(t) of Figure 9.5, the greater acceleration in the up motion is visible. The common learning obstacle that in the up motion, acceleration is null, is overcome by analyzing and correlating the three kinematical graphs.

Figure 9.8 shows how the acceleration due to the friction force causes different values of total acceleration in the up/down motion.

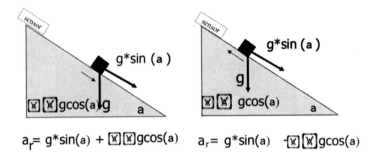

$$a_r = g*sin(a) + \boxed{x}\boxed{x}gcos(a) \qquad a_r = g*sin(a) - \boxed{x}\boxed{x}gcos(a)$$

Figure 9.8 Vector schema of the total acceleration of a cart moving up/down a ramp.

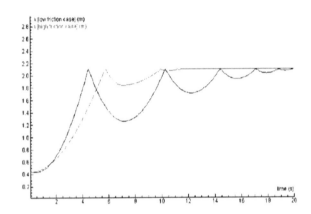

Figure 9.9 Cart up/down a ramp (sensor at ramp top). (a) medium friction, (b) higher friction. The cart starts moving down, hits an obstacle at the ramp bottom, goes up again, and so on until it stops.

The effects of increasing friction can be easily studied by adding to the cart a strip sliding on the ramp. To study multiple up/down, a cart with a spring hitting a rigid obstacle at the ramp bottom is useful (Figure 9.9).

The effects of increased friction are clearly visible: the medium friction s(t) shows five full (quasi) parabolas, less and less deep; the greater friction shows two full (quasi) parabolas before the cart stops at the ramp bottom; the symmetry of the parabolas is altered. A cart up/down motion is linkable with a multiple bouncing of a rubber ball on the floor; the ball's (quasi) free fall being conceptually equivalent to a low-friction cart motion on a ramp (Figure 9.10).

The dissipative effects are clearly visible in the ball s(t): the parabola depth decreases with the number of bounces as seen in the velocity values after hitting the floor and the duration of the constant acceleration motion. In

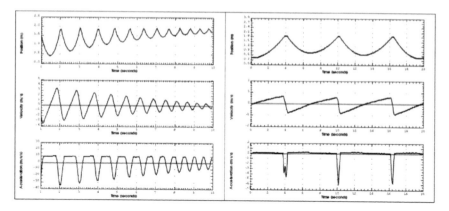

Figure 9.10 At right, the up/down motion of a cart (three fast kicks to the cart when at the ramp bottom). At left, about 12 bounces of a rubber ball on the floor.

the cart's v(t), the transitions from negative to positive values, which are not step-like, indicate the impossibility of infinite acceleration in a physical phenomenon; the learners are helped to differentiate a real motion v(t) from a mathematical step function. All these activities refer to different facets of the same phenomenon and therefore belong to the area named *topical knowledge*.

Networked knowledge can be linked to mathematics-physics relationships; best fit procedures; experimental abilities (increasing friction, measuring multiple bouncing, etc.). For instance, the peaks in a(t) allow pointing/linking to the impulsive forces acting in the cart collisions at the ramp bottom and in the ball collisions with the floor. When a force sensor is on the cart, the time trend of a force F(t) is also measurable; the very short duration of the impulsive forces in a collision can be correlated with the a(t) peaks. To make "visible" forces that are usually "invisible" opens new ways to teach/learn the third law of dynamics (see later) and overcomes common robust learning obstacles in understanding its meaning and going beyond the "action equals reaction" slogan.

From the viewpoint of *meta-knowledge*, this example fosters reflections on the complex path required to reach the ideal case laws; on how to bridge the kinematic viewpoint with the dynamic one (usually taught/perceived as unrelated); on differences among phenomenological forces and universal interactions (gravitational, electromagnetic, strong and weak); on descriptive power, validity range and limits of models; etc.

GALILEO'S COMPOSITION OF VELOCITY

The classical, prerelativistic composition of velocity is usually taught by formulas about vector relationships in linear transformations of coordinate systems in relative motion with constant velocity. Very often (especially at secondary school), this presentation is not effective, and several learning obstacles are encountered. The path presented here roots Galileo's composition of velocity in perceptual knowledge by exploiting an RTEI OLE approach, which is very similar to the cart-ramp example (SECIF-KINFOR, 2004).

The Prediction phase elicits ideas/reasoning about velocity composition in familiar experiences and doable experiments. The Experiment phase firstly explores the motion of an object moved back and forth by a still person extending/retracting rhythmically his/her arms. Then the person walks very regularly, moving the object as before. The analysis focuses on s(t) and v(t); each measure lasting less than a minute and each learner can experiment and integrate her/his perceptual knowledge with the abstract representation of kinematical graphs (Figure 9.11). The more rhythmic the motion of the arms, the more regular the s(t) and v(t) trends.

Two motions are combined: the walk of the person (quasi constant velocity, positive or negative) and the "back/forth oscillations" of the book. The regular walk produces the average slope of s(t), negative (positive) in the left (right) graph because the person is moving toward (away from) the sensor. The v(t) shows clearly how the walk velocity, with its sign, combines with the book velocity; it is easy to infer the phenomenological rule "the velocity of the book moved by the person is the algebraic sum of the velocity of the walking person and the velocity of the book moved by a still person." Further, study a person moving the book while seated on a wheeled office chair moved rhythmically back and forth by another person, in phase or counter-phase with the book motion; the goals are to explore another situation well understandable in terms of perceptual knowledge and to check if the rule

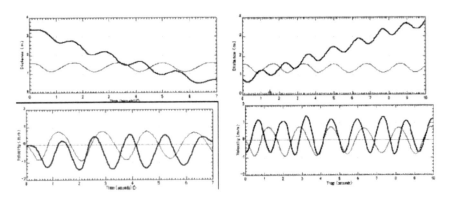

Figure 9.11 A person moves a book rhythmically by extending/retracting the arms. Light curves refer to a still person, dark curves to a person walking regularly toward (left graphs) and away (right graphs) from the sensor. The same time interval is represented in the left (right) graphs to ease correlating s(t) and v(t).

found can be generalized to reinforce the plausibility of the Galileo vector composition of velocity. These activities also refer to the *topical knowledge* in terms of overcoming learning difficulties and facilitating the development of formal thinking about the velocity composition.

In terms of *networked knowledge,* experimental abilities are acquired in the in-phase or counterphase motion, historical reflections link with physics at Galileo times, and advantages of vectors' formalism are discussable. A reflection about harmonic oscillations is often triggered by the *"oscillatory"* shape of the book, s(t). The learners' attention is often captured mainly by the graph shape that acts as a Gestalt, obscuring information conveyed by the phenomenon description, the axes variables, or the image caption. This is addressable by comparing the kinematical graphs of a person who stands still and rocks back/forth on his/her feet as rhythmically as possible with the (quasi) harmonic oscillation of a spring-mass system (Figure 9.12)

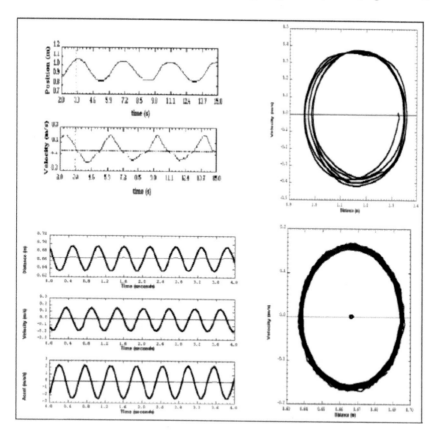

Figure 9.12 Rhythmic rocking on feet of a standing person (top), quasi harmonic oscillations of a spring-mass system (bottom).

The rocking person, s(t), easily evokes a (quasi) harmonic trend in time; the corresponding v(t) is clearly nonharmonic. The v(s) graphs allow the introduction of the concept of phase space. The spring-mass system oscillations are (quasi) free harmonic ones (some air resistance is unavoidable as some friction at the suspension); s(t), v(t), and a(t) are good approximations of harmonic functions, the "thickness" of v(s) indicates some damping due to the above dissipative effects. The fact that similar or very similar shapes do not imply the same dynamics can be clarified. Furthermore, a bridge with the elastic force determining the spring-mass oscillations and its simple formal expression can be proposed; the force acting in the rocking motion being of diverse nature and very complex formalization.

Several aspects refer to *meta-knowledge*. First, to reflect on the main features of modeling, as the same law or model can represent very different phenomena, according to the variables' meaning. For instance, gravitational and electric forces, two out of the four fundamental interactions governing our natural universe, have an inverse square dependence on the distance between point-like masses and charges. An exponential model (increasing or decreasing) describes phenomena as diverse as radioactive decay, growth and death of some species, etc. Second, to reflect on the descriptive power and validity range of models and paradigms, Galileo's velocity composition (a pillar of Classical Mechanics) is an approximation of the relativistic composition of velocity. Naturally comes a pointer to the Special Relativity theory by Albert Einstein, one of the most significant changes of paradigm, a crucial conceptual and historical development of 20th-century physics.

FORCES EXPLORATION

The last emblematic examples are briefly discussed, mainly to suggest activities based on using a force sensor.[4] These activities are useful for at least two purposes: easing the study of forces and establishing bridges with physics topics where forces intervene. A nonexhaustive list of examples follows.

Perception and Common Language About Force

In commonsense knowledge, the terms pull and push are related to the intuitive idea of force as well as to the perceptual awareness of some effort. In everyday language, the term force is also used to mean energy, sustained effort, capability of doing work, etc. Often the learners' naïve ideas about force conflict with its disciplinary definition, the main ambiguity referring to force and energy. The perceptual knowledge of a pull/push on the force sensor correlated with the measured F(t) eases the comprehension of what a force is and of the conventions used for the force sign. Comparing experiments using a standard dynamometer for static measures with those using a force sensor (several tens of data/second) allows the appreciation of the study of the time trend of a force and the meaning of the details observable.

Elastic Forces

Elastic bodies are commonly used in everyday life, but naïve ideas/reasoning about elastic properties, why a deformed elastic object tends to return to its in-deformed state, why it sometimes breaks, etc. are often unclear, ambiguous, and conflicting with physics knowledge. Experiments using a force sensor and exploring the behavior of diverse types of elastic and non-elastic objects (e.g. rubber bands, elastic fabrics, metallic springs, pieces of string, electric wires, etc.) allow to visualize and measure the static force producing a certain deformation (if any); measure the elastic constant; analyze if a linear trend is present when applying a force increasing with time; and measure the force at the break limit. The learning difficulties about the minus sign in the Hook law can be elicited, and the intrinsic meaning of this sign clarified. In experiments about oscillating systems (e.g., a suspended metallic spring or rubber band, plus a weight), the elastic force can be explored in a dynamic case. This study allows also to address the naïve reasoning that *always* assumes a force parallel to velocity when a system moves: the elastic force at play in the harmonic oscillator is always opposite to the spring deformation and there is not a force in the same direction of the velocity; and the gravitational force determines only the equilibrium position of the spring-mass system.

The Friction Force

In everyday life, the effects of friction are experienced very often, for example, when walking, driving a vehicle, cleaning a glass, holding a bottle, rubbing one's own hands, etc.

However many learners encounter difficulties in explaining what friction is. Often vague ideas are expressed about something contrasting the "motion" or making people slide, or lighting a match. Experiments using a force sensor can help learners to understand that friction is a phenomenological force whose analytic expression depends on the particular context. Measurements of the time trend $F(t)$ of the forces needed to move a block pulled by hand via a force sensor on different types of surfaces and of those needed to move blocks of different materials on the same surface help to understand that the friction force depends on a local interaction of body-

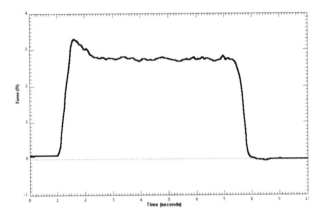

Figure 9.13   Time graph of the force needed to move a still cart on a smooth track and to maintain it in motion with quasi constant speed.

surface and the characteristics of the surfaces in contact. The difference between the static friction coefficient and the dynamic friction one is clarified by studying how $F(t)$ changes when a still body pulled by hand via a force sensor starts to move. Two diverse values are measured, greater at the verge of motion and smaller (quasi constant) to keep the body in motion with constant velocity (Figure 9.13).

Forces in a Collision

In studying the physics of collisions, several learning difficulties come from the fact that the forces acting during the phenomenon are "visible" only in terms of what happens as a result of the collision. The standard teaching focuses on a pre/post description of trajectories, momentum, and energy of the colliding bodies. The definition of elastic collision refers to situations far from common experience. Experiments about collisions between two bodies on which force probes are mounted allow to "see" the impulsive forces acting during the very brief hit and, by analyzing the $F(t)$ graphs, measure their intensity, direction, and duration. This approach clarifies operationally how to distinguish an impulsive force from a constant one (e.g., a weight suspended from a force sensor, a constant pressure exerted by hand on the sensible part of the sensor, etc.). The analysis of the $F(t)$ in a collision offers information that helps to understand the third law of dynamics. The common naïve and incorrect idea that in a collision the heavier or bigger object exerts a greater force than the lighter or smaller one can be addressed by experimenting with two carts of diverse weight. The forces exerted during the collision by each cart have equal intensity

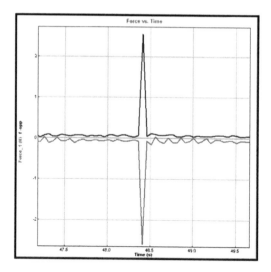

Figure 9.14   F(t) graphs in the case of a head-on collision of two carts of different weight.

and opposite direction (Figure 9.14). The collision has lasted a few tenths of a second, the (quasi) zero values of the forces pre/post the collision indicate that the carts moved on a low-friction track.

Force and Motion

An educationally important synergy results from the simultaneous use of a motion and a force sensor. Their independent measures allow us to correlate kinematical measurements as s(t), v(t), and a(t) with a dynamical viewpoint, in particular, the force on the moving object. Experiments using the two sensors and exploring different motions can help understand the second law of dynamics, perceived by many learners as very abstract, difficult, and far from common experience. To measure that F(t) and a(t) have the same time trend allows reflection on the relationships between force and acceleration. The experimental study of a spring-mass system, suspended from a force sensor, whose one-dimensional oscillations are measured by a motion sensor, is particularly useful in studying the correlation between a(t) and F(t). This type of measurement can help learners understand facets of what **F** = m **a** means and suggest reflections about the long and complex process from the Aristotelian connection force-velocity to the Newtonian force-acceleration.

All the examples discussed address mechanics content; this choice is due to the fact that they are presented whenever physics is taught. Moreover, in all of them, the RTEI OLE approach builds on the connection between perceptual knowledge and abstract (graphs) representations of data; this can be an important help for those learners whose capability of abstraction is not yet fully developed.

CONCLUSIONS

Labwork, when practiced with disciplinary competence, pedagogical wisdom, and students' participative involvement, can significantly change the quality of physics education. RTEI further empowers the potential of labwork and represents a typical case of mutual fertilization between educational practices and technological opportunities; a combination of perceived educational needs, pedagogical creativity, and technical competencies have been brought to the creation of RTEI systems and, on the other side, sensors online with the computer that have further opened the doors to new learning opportunities.

An appropriate use of RTEI can have a great impact in terms of educational descriptors as Pedagogical Effectiveness (PE) and Transformation Potential (TP), as suggested in the above examples. These descriptors, useful for many educational methods, approaches, and materials are very appropriate for RTEI (Feiner-Valkier & Sassi, 2010).

Pedagogical Effectiveness has to do with the capability of a specific didactic approach to be a vector of meaningful *pedagogical* innovation. Actually, RTEI OLEs encourage learners' centered and inquiry-based learning; bridge phenomenological data with their abstract (and formal) representations and theoretical reflections with concrete technical operation; help establish links among different content areas; and enhance learners' interest and motivation.

Transformation Potential refers to the capability of a didactic approach to be a vector of *cultural* innovation for both teachers and learners. On the teachers' side, RTEI OLEs can contribute to transforming conceptions about teaching and learning; to establishing new attitudes and interpretations of the role of the teacher in PE; to integrating informal and nonformal learning in school education; and to building the foundations of a new pedagogical culture in PE. On the learners' side, RTEI-based approaches can contribute to changing beliefs about how to learn; fostering constructive interactions with peers and teachers; and improving the awareness of the scientific methods and of image of physics and, more generally, of science.

RTEI-based approaches offer a specific contribution to the quality of physics education: the relations between specific phenomena and related laws (*topical knowledge*) can be more easily understood and interiorized; the theoretical and the operational levels can be easily bridged; and the comprehension of the connections among different, but related, physical topics (*networked knowledge*) can be fostered. And last but not least, the use of RTEI as Open Environments favors the gradual development of the ability to build new physical knowledge (*meta-knowledge*). In other terms, RTEI may help to replace the traditional outcome of physics education, for example, a superficial knowledge that is easily forgotten, with a well-rooted knowledge capable of further autonomous expansions and useful as a basis for a critical understanding of new scientific developments.

Despite their powerful features, RTEI OLE approaches are not yet commonly adopted. The reason is mainly cultural since the cost of equipment, which in the past was considered one of the causes of the slow diffusion of RTEI, is nowadays quite affordable, at least in the industrialized countries.

The naturalization of RTEI approaches requires of the teacher to change well-established teaching habits and to integrate his disciplinary and pedagogical knowledge with new attitudes and new abilities: switching from a transmissive to a constructive approach; integrating theoretical reflection and experimental activities; dynamically shaping the teaching/learning process together with the learners; adopting a maieutic approach where the, possibly naïve, learners' ideas can emerge and become focal points of the learning process; etc. Teachers seldom receive adequate support since RTEI OLE are not commonly proposed in preservice and in-service education, coverage in textbooks is scanty, and suggestions for designing and conducting effective activities as well as research. All these factor originate the most common problems in adopting RTEI approaches.

A process of cultural growth, by its very nature, cannot happen all of a sudden as a result of applying specific recipes. This holds also for the naturalization of RTEI approaches which, rather than applying recipes, requires us to create the conditions that motivate, favor, and support that change. Here it is worth mentioning at least three conditions that appear necessary for this cultural transformation:

- Support teachers to build a sound disciplinary knowledge:
  RTEI approaches require travelling from ideas to experiments and vice versa, dealing with unforeseen situations, suggesting experimental ways that may take the learners from naïve conceptions to physical knowledge. This requires a mastery of the discipline in its theoretical and experimental aspects. Moreover, the existence of grey areas of blurred knowledge that needs to be reconstructed can be scary for teachers who want to change their teaching approach.
- Make available to teachers meaningful educational and didactic resources:
  Diverse types of support resources can help teachers willing to approach RTEI or to enhance their present experiences. Different types of resources may be of interest: descriptions of commented

RTEI OLE activities, experimented in ordinary contexts and enriched with teachers' and learners' reactions; examples of problems encountered by fellow teachers together with their practical solutions; example proposals discussed in terms of *networked knowledge* and *meta-knowledge*; reflections about which changes in teaching/learning occurred and how they were detected

- Fostering sharing and collaboration among teachers and learners: Each individual has a different patrimony of experiences and intellectual and emotional resources. Educational research has shown that sharing and collaboration among teachers and learners has several advantages: it enhances individual points of views; prevents difficulties and errors that someone in the group has already experienced; and from the psychological point of view, represents a form of co-counseling and a factor-enhancing responsibility and motivation.

ACKNOWLEDGMENTS

Several of the proposals and materials discussed have been researched and tested by the DF/ICT research group of the Physics Department of Università "Federico II," Napoli, Italy. Many thanks to Sara Lombardi, Gabriella Monroy, and Italo Testa.

NOTES

1. Learners means students enrolled in formal education, preservice and in-service teachers attending teacher education programs, those who learn in nonformal and informal educational environments, and anyone who learns in any learning situation.
2. Physical Science Study Committee.
3. Several cycles referring to the same phases have been proposed, e.g., POE (Predict, Observe, Explain) suggested when simulations are used; POEC (Predict, Observe, Explain, Compare); OPCE (Observe, Predict, Compare, Explain) when experiments are made. In the latter, Predict is the second phase because the prediction can also be related to the simulation/model describing the phenomenon observed.
4. One type of force sensor is based on the Hall effect. Preliminarily, it is needed to become familiar with the force probe, its calibration procedure, and the zero force level.

REFERENCES

Champagne, A. B., Gunstone, R. F., & Klopfer, L. E. (1985). Effecting changes in cognitive structures among physics students. In H. T. West & A. L. Pines (Eds.), *Cognitive structure and conceptual change* (pp. 61–90).

Chinn, C. A., & Brewer, W. F. (1993). The role of anomalous data in knowledge acquisition: A theoretical framework and implications for science instruction. *Review of Educational Research, 63,* 1–49.

Duit, R. (2009, March). Bibliography – *STCSE (Students' and Teachers' Conceptions and Science Education)* Retrieved July 27, 2010, from http://www.ipn.uni-kiel.de/aktuell/stcse/stcse.html

Feiner-Valkier, S., & Sassi, E. (2010). On the use of new methods and multimedia. *Il Nuovo Cimento, 33*(C), 3.

French, A. P. (1986). Setting new directions in physics teaching: PSSC 30 years later. *Physics Today, 39*(9), 30.

Hofstein, A., & Lunetta, V. (1982). The role of the laboratory in science teaching: Neglected aspects of research. *Review of Eduational Research, 52*(2), 201–217.

Hofstein, A., & Lunetta, V. (2004). The laboratory in science education: Foundations for the twenty-first century. *Science Education, 88*(1), 28–54.

Linn, M. C., & Eylon, B-S. (2006). Science education: Integrating views of learning and instruction. In P. A. Alexander & P. H. Winne (Eds.), *Handbook of educational psychology* (2nd ed.). Mahwah, NJ: Lawrence Erlbaum Associates.

Maddox, J. (1966). The Nuffield Physics Project. *Physics Education, 1*(3).

Pinto, R., Ametller, J., Couso, D., Sassi, E., Monroy, G., Testa, I., & Lombardi, S. (2003). Some problems encountered in the introduction of innovations in secondary school science education and suggestions for overcoming them. *Mediterranean Journal of Educational Studies, 8*(1), 113–134.

Rieber, R., & Carton, A. (Eds.). (1987). *The collected works of L. S. Vygotsky* (Vol. 1). New York: Plenum Press.

Sassi, E. (2002). *Real-time approaches in the development of formal thinking in physics.* In proceedings of the First International GIREP Seminar 2001. "Developing Formal Thinking in Physics" (M. Michelini & M. Cobal, Eds.). Forum Editrice Udine, pp. 40–51.

Sassi, E. (2005). *Some views about research in physics education.* Proceedings of 1st European Physics Education Conference, Bad Honnef, Germany. Retrieved September 14, 2010, from http://www.physik.uni-mainz.de/lehramt/epec/sassi.pdf

Sassi, E., & Vicentini, M. (2008). Aims and strategies of laboratory work. In M. Vicentini & E. Sassi (Eds.), *Connecting research in physics education with teacher education* (Vol. 2). International Commission on Physics Education.

SECIF-KINFOR. (2004). *Teaching and learning about kinematics and force.* (in Italian, http://www.fisica.unina.it/Gener/did/kinfor/SeCif/cover.htm)

Séré, M-G., Leach, J., Niedderer, H., Psillos, D., Tiberghien, A., & Vicentini, M. (1996, February–1998, April). Final Report: Improving Science Education: Issues and Research on Innovative Empirical and Computer-based Approaches to Labwork in Europe. *European Commsion, Targeted Socio-Economic Research Programme, Project PL 95-2005, Labwork in Science Education.* http://www.ictt.insa-lyon.fr/ELabs/Bibliographie/e-learning/Rapports/CEE-LABWORK%20 IN%20SCIENCE%20EDUCATION.pdf

Sjøberg, S., & Schreiner, C. (2010, March). *The ROSE project: An overview and key findings*. Oslo, Norway: University of Oslo.

Testa, I., Monroy, G., & Sassi, E. (2002). Students' reading images in kinematics: The case of real-time graphs. *International Journal of Science Education, 24*(3), 235–256.

Thornton, R. (1987). Tools for scientific thinking – Microcomputer-based laboratories for physics. *Physics Education, 22*(4), 230.

Thornton, R. K. (2008). Effective learning environments for computer supported instruction in the physics classroom and laboratory. In M. Vicentini & E. Sassi (Eds.), *Connecting research in physics education with teacher education* (Vol. 2). International Commission on Physics Education.

Thornton, R., & Sokoloff, D. (1998). Assessing student learning of Newton's laws: The force and motion conceptual evaluation and the evaluation of active learning laboratory and lecture curricula. *American Journal of Physics, 66*, 338–352.

Trumper, R., (2003). The physics laboratory – A historical overview and future perspectives *Science & Education, 12*, 645–670.

Viennot, L. (2003a). *Reasoning in physics*. Dordrecht, The Netherlands: Kluwer.

Viennot, L. (2003b). *Teaching physics*. Dordrecht, The Netherlands: Kluwer.

Viennot, L. (2006). Teaching rituals and students' intellectual satisfaction. *Physics Education, 41*(5), 400–408.

White, R. T., & Gunstone, R. F. (1992). *Probing understanding*. London: Falmer Press

CHAPTER 10

ENHANCING ASYNCHRONOUS LEARNING IN A BLENDED LEARNING ENVIRONMENT

Mun Fie Tsoi
National Institute of Education, Singapore
Nanyang Technological University

ABSTRACT

Research on asynchronous learning in a blended learning environment has led to a variety of pedagogical practices. As such, this chapter provides an evidence-based research practice model, namely, the TSOI Hybrid Learning Model, to enhance asynchronous learning, a fundamental component present in blended learning. The hybrid learning model is advanced from the Science learning cycle model and the Kolb's experiential learning cycle model.

This hybrid learning model, inclined toward constructivism, represents learning as a cognitive cyclical process of four phases: **T**ranslating, **S**culpting, **O**perationalizing, and **I**ntegrating. A major feature is to promote active cognitive processing and addressing learning style. An application is illustrated with an authentic example on understanding multimedia learning in science education for preservice teachers of the science degree course. The hybrid learning

Contemporary Science Teaching Approaches, pages 213–228

213

model guides the blended learning design involving asynchronous learning using Web 2.0. Learning outcomes in terms of richness of online discussions, thinking processes, and personal reflections have been positive. Implications will be discussed in the context of Web 2.0 in science education.

INTRODUCTION

Research on the nature of asynchronous learning and its features in a blended learning environment has led to a variety of pedagogical practices. Asynchronous learning provides the learner the opportunity to interact online with peers or the teacher not concurrently, thus allowing reflective time. This chapter sets out an evidence-based research practice learning model as a practice-based framework for enhancing asynchronous learning, a fundamental component present in blended learning. As in previous research studies on its functions and applications conducted by Tsoi (2007, 2008, 2009), in which the learning and affective outcomes have been positive, the hybrid learning model will also contribute as an innovative framework for enhancing asynchronous learning in blended learning.

THEORETICAL FRAMEWORK OF THE TSOI HYBRID LEARNING MODEL (TSOI HLM)

The theoretical framework of the TSOI HLM is evolved from the Science learning cycle model and the Kolb's experiential learning cycle model. The term hybrid means the mixing of two different things to give a better product, which in this case, is a learning model that is pedagogically more innovative and comprehensive than each of the original models, namely, the Science Learning Cycle model and the Kolb's Experiential Learning Cycle model. The inquiry-based Science learningcycle model represents an inductive application of information-processing models of teaching and learning (Karplus, 1977; Lawson, 1995; Renner & Marek, 1990). It has three phases in a cycle: exploration, concept invention, and concept application, as shown in Figure 10.1. The exploration phase focuses on "What did you do?" while the concept invention phase places emphasis on "What did you find out?" The concept application phase requires the application of the concept.

The exploration phase (gathering of data) is often accomplished during a science activity or an experiment. During this exploration phase, learners learn through their own actions and reactions in a new situation and have the opportunity to explore new learning materials and new ideas with minimal guidance from the teacher.

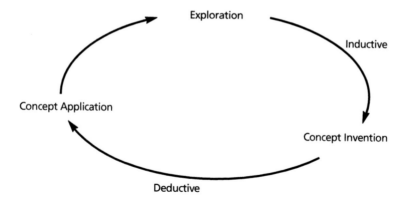

Figure 10.1 The Science learning cycle model.

The concept invention phase gives the opportunity to the student and/ or teacher to derive the concept from the data through classroom discussion. This phase involves the introduction of a new term or terms. Ideally, learners are encouraged to discover as much of a new pattern as possible before the term is revealed to them. The third phase, concept application, allows the student to explore the relevance and application. In this last phase, the learners apply the new term(s) to additional problems. The concept application phase is essential, as it allows learners to extend the range of applicability of the new concept.

The Kolb's experiential learning cycle (Kolb, 1984), as shown in Figure 10.2, represents learning as a process in a cycle of four stages, namely, concrete experience, reflective observation, abstract conceptualization, and ac-

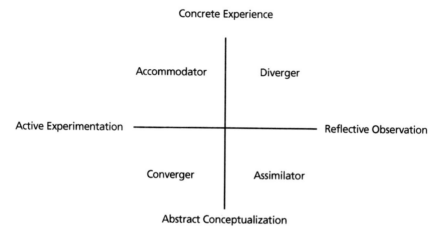

Figure 10.2 Kolb's experiential learning model (Smith & Kolb, 1986, p. 16).

tive experimentation. The concrete experience stage is about "doing," while the reflective observation stage concerns "understanding the doing." The abstract conceptualization stage focuses on the "understanding" part, and the active experimentation stage is about "doing the understanding." The core idea in the Kolb's experiential learning cycle is that learning requires both a grasp or figurative representation of experiences and some transformation of that representation. This experiential learning cycle model has also been used as a framework for organizing interactive multimedia learning activities (His & Agogino, 1994; Tsoi & Goh, 1999; Van Aalst, Carey, & McKerlie, 1995).

Kolb also created four quadrants in his model of experiential learning. He assigned each quadrant a learning style: diverger, converger, assimilator, and accommodator (see Figure 10.2).

For convergers, experience is grasped through abstract comprehension and transformed through action, which combines abstract conceptualization and active experimentation. For divergers, experience is grasped as opposite of convergers, that is, concretely through feelings and transformed through thought, which combines concrete experience with reflective observation. For assimilators, experience is grasped through abstract comprehension and transformed through thought, which combines abstract conceptualization and reflective observation. For the accommodators, experience is grasped concretely through feelings and transformed by action, combining the features of concrete experience and active experimentation.

The TSOI Hybrid Learning Model conceptualized by Tsoi (2007) represents learning as a cognitive process in a cycle of four phases: Translating, Sculpting, Operationalizing, and Integrating. The foremost feature is to promote active cognitive processing in the learner for meaningful and engaged learning, proceeding from inductive to deductive learning. Besides, it is inclined toward constructivism. Figure 10.3 shows the four phases of this learning model.

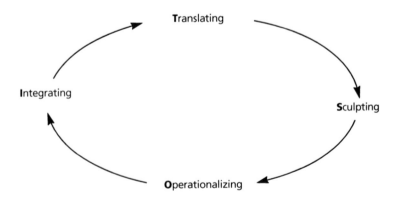

Figure 10.3 TSOI hybrid learning model.

The Translating phase is similar to the exploration phase of the Science learning cycle model and the concrete experience stage of Kolb's experiential learning cycle model. This is where interactive experiences are translated to beginning ideas or concepts to be further engaged in the Sculpting phase. The Sculpting phase parallels the concept invention phase of the Science learning cycle model and predominantly the reflective observation stage of the Kolb's experiential learning cycle, including partially the abstract conceptualization stage of the Kolb's experiential learning cycle. This is where the beginning idea or concept, still in its raw form, is further molded into a concrete form that is meaningful to the learner.

The Operationalizing phase is predominantly similar to the abstract conceptualization stage of the Kolb's experiential learning cycle and involves increasing the understanding of the relationship between thinking and concept acquisition. The Integrating phase parallels the concept application of the Science learning cycle model as well as the active experimentation stage of Kolb's experiential learning cycle. This is where the concept is applied to new domains in which the transfer of learning is practiced.

In pedagogical practice, the Translating phase emphasizes initial concept exposure for preliminary experience. The instructional learning activity, though general in nature, is designed to have an initial relationship to the principle underlying the concept, which is to be further engaged in the second phase, the Sculpting phase. The Sculpting phase emphasizes concept construction for its critical attributes. The concept, still in its beginning or raw form as taken from the Translating phase, is logically sculpted or shaped to a more concrete form by a series of appropriate and relevant instructional learning activities that are designed meaningfully to assist the learner to identify the critical attributes of the concept.

The Operationalizing phase emphasizes concept internalization for its meaningful functionality. A more scientific view of the concept is formed and internalized for meaningful functionality. This important phase is crucial as it serves as the vital bridge connecting the Sculpting phase and the Integrating phase for not only concept formation but also concept internalization in which all the critical attributes of the concept are linked together so as to prepare the learner to be operationally ready for further applications in the Integrating phase. During the fourth phase, the Integrating phase, the just-learned concept is applied to new situations as well as integrated into different contexts in order for meaningful learning to occur. The Integrating phase emphasizes concept application for meaningful transfer of knowledge.

Blended Learning Design for Face-to-Face Interactions (Translating Phase)

The hybrid learning model guides the learning design for enhancing asynchronous learning in a blended learning environment. An example of a curriculum studies course on understanding multimedia learning pedagogy in chemistry education for 25 pre-ervice teachers of the BSc/BA (Ed) course (Bachelor of Science/ Bachelor of Arts (Education), Secondary) is illustrated. In the Translating phase, face-to-face interactions infused with cooperative learning approaches such as roundtable, think-pair-share, and number heads together are carried out in three sessions of 2 hours each with group discussions. For example, responses to a question on "what do you understand by the term multimedia learning" are elicited and built upon for further discussion. The main idea is to provide a preliminary experience as to what the term multimedia learning means to them. After which, the discussed ideas and points are linked to the next main activity, which focuses on the fundamentals of TSOI HLM as the pedagogic model for multimedia learning design to prepare the trainee teachers for the Sculpting phase. As such, the beginning idea of the TSOI HLM will be further engaged in the Sculpting phase. The Translating phase emphasizes initial concept exposure for preliminary experience. The instructional learning activity, though general in nature, is designed to have an initial relationship to the principle underlying the concept, which is to be further engaged in the second phase, the Sculpting phase. In chemistry education, stoichiometry, an abstract and difficult topic, is used (Tsoi, Goh, & Chia, 1998) to illustrate the understanding and applications of the hybrid learning model. One of the subtopics used is solution concentration. This next section will provide insights on the design application of the TSOI HLM in multimedia learning.

Blended Learning Design for Asynchronous Learning (Sculpting Phase)

Basically, for this second phase, the Sculpting phase, there are two e-learning components to complete in the learning content management system, namely, multimedia learning and asynchronous learning. During this phase, knowledge of the concept is beginning to be constructed based on the learner's facilitated classroom experiences from the Translating phase and the learner's guided multimedia learning experiences as well as asynchronous learning experiences from the Sculpting phase. The concept, still in its beginning or raw form as taken from the Translating phase, is logically sculpted or shaped to a more concrete form by a series of appropriate and relevant instructional learning activities that are designed meaningfully to assist the learner to identify the critical features of the concept, which in this case, is the TSOI HLM.

Multimedia learning is done in the form of a stoichiometry (solution concentration) multimedia module. Asynchronous learning is achieved via online discussion board. The solution concentration multimedia module consists of four instructional learning episodes in accordance with the four phases of the TSOI HLM. These four instructional learning episodes are (a) Which is more concentrated, (b) Physical Meaning and Definition, (c) Investigating Chemical Reactions, and (d) A Simple Equation. The Translating phase of the TSOI HLM is illustrated to show a portion of the learning model, since the focus of this chapter is not on designing a multimedia learning module. The first instructional learning episode, "Which is more concentrated," as a Translating phase, provides the learner with visual representations in which the relationship between the concentration of the solution and the amount of solute particles dissolved in the solution will be formulated cognitively. In the first activity, the learner is asked to compare two solutions in terms of concentration, which involves the amount of solute particles dissolved in the solution (see Figure 10.4). Observational response is elicited. This is then extended to the second activity where the learner is posed a question as to how to make the concentration of solution A the same as solution B (see Figure 10.5). These visual representations

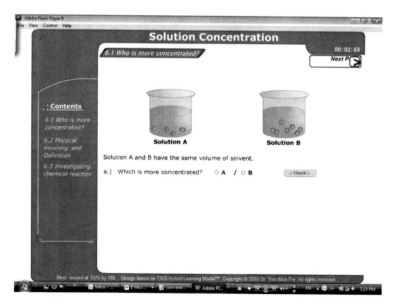

Figure 10.4 Solution concentration multimedia learning module (Translating phase).

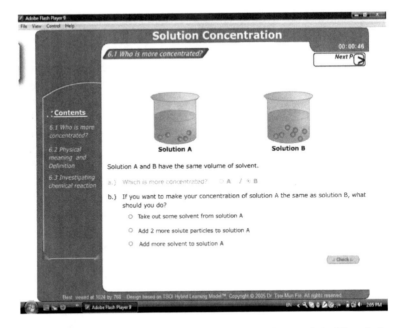

Figure 10.5 Solution concentration multimedia learning module (Translating phase).

serve as a beginning idea or concept of solution concentration, which is further molded in the next Sculpting phase in the multimedia learning module, Physical Meaning and Definition.

A week is given to the individual learner to perform the task of observing how the multimedia learning module is developed based on the TSOI HLM as its pedagogic design model and reflecting on one's observations. The duration of the module is approximately 45 minutes. Relevant resources such as appropriate journal articles and course PowerPoint slides are provided for a meaningful reflection.

The following week involves the group members going online using the discussion board to express their thoughts and ideas and discuss their observations. The online discussion board is used not only for its ease in providing comments but also as a platform for encouraging and enhancing asynchronous learning. Relevant questions for group discussion provided are, for example, observe the way the concept of concentration is built up and learned using this hybrid learning model; the multiple multimedia representations used; the kind of questions posed, the type of examples used, the variety of practice problems embedded; the macroscopic, submicroscopic, and symbolic levels used; and your impressions or feelings of this hybrid learning model utilized in the design of this stoichiometry multimedia learning module. Primarily, a fundamental understanding of the hybrid learning model in terms of its critical features is constructed within these 2 weeks of multimedia learning as well as asynchronous learning. Samples of asynchronous learning are shown (see Figures 10.6, 10.7, 10.8, 10.9, and 10.10).

Thread: Concept development of concentration Post: Concept development of concentration Author: CHIN HOOI MING (BS20607)	Posted Date: January 19, 2010 8:53 AM Status: Published

From this learning package, we can see how the concept of concentration has been developed in a systematic way. It starts of by asking students the qualitative way to differentiate concentrated and diluted solution (section 6.1), which is important in the 'Translating' phase of the TSOI learning model. Then the concept of concentration is further constructed by showing the qualitative methods to calculate concentration. The calculation for number of moles also has been introduced in a logical order-formula for number of moles in liquid and solid in section 6.2, then gas in section 6.3. These are very important in the 'sculpting' phase.

I like the idea of giving some space for students to write down their own understanding of the concept in the 'self-explanation' column in section 6.3. I think this will help the students to internalize what they have learnt, before bringing the students further to apply the learnt concept in various examples in section 6.4.

Thread: Concept development of concentration Post: RE: Concept development of concentration Author: CHER CHUIN YUAN (BS20607)	Posted Date: January 20, 2010 11:10 AM Status: Published

I agree with Hooi Ming. I find that for a lot of students, they can understand a concept but experience difficulty putting it in words. Therefore, giving them an opportunity to explain a concept in their own words is a good idea.

In this way students can "struggle" with the expression of their ideas, and with the teacher's comments, improve on how to bring their idea across to others. After all, in science, the most important thing is not to discover knowledge, but to express that knowledge to others. This will give them a good foundation into being a good scientist.

Thread: Concept development of concentration Post: RE: Concept development of concentration Author: LIM ELEEN (BS20607)	Posted Date: January 23, 2010 2:32 PM Status: Published

In Section 6.3, students get to internalize their mass solution stoichiometry concepts through examples and practice. I agree with Hooi Ming and Daniel that through self explanation, students to understand what they have learnt. They would have to ask themselves why they had to carry out the certain calculation or step in order to attain the answer in the question.

I think that there are also sufficient practices in section 6.3 for students to understand the stoichiometry concepts. The exercises and calculations were taught stepwise. Students would have click on the "continue" or orange button in order to move on to the next step. This enables them to learn and understand the calculations at their own pace. They are hence not pressurized and are less likely to feel confused over a certain step.

Figure 10.6 Discussion board (concept development of concentration).

Thread: Systematic Model Restricts Creativity?
Post: Systematic Model Restricts Creativity?
Author: CHER CHUIN YUAN (BS20607)

Posted Date: January 22, 2010
10:38 PM
Status: Published

I was reading some of the comments in the other discussion group and I found this point mentioned by Bryan to be quite interesting. He mentions that the Hybrid Learning Module is good for its systematic approach, but he is concerned that it will restrict students from expressing creativity.

I feel Bryan raises a good point, although I disagree with him. As evident from our presentations on Thursday, most of us would conduct a guided inquiry-styled lesson if we are planning a lesson based on this model. This would give students space to think and to be creative. If the lesson is structured to make the students think out of the box, I don't see how the model will be restricting creativity.

As such, although I feel there is no need for undue worries, we will also have to take note on how we are planning our lessons to make sure students are given space to express their creativity.

In addition, from what I understand about the model, it is not only systematic, but also very flexible. For example, in our lesson on Thursday, Dr. Tsoi mentioned that strategies applied in the "Sculpting" stage can also be used in the "Translation" stage and vice versa. So it really depends on the teacher's rationale and judgement when it comes to planning and carrying out a creative lesson.

What are your views?

Thread: Systematic Model Restricts Creativity?
Post: RE: Systematic Model Restricts Creativity?
Author: LIM ELEEN (BS20607)

Posted Date: January 23, 2010
3:20 PM
Status: Published

I agree with Daniel that although the Hybrid Learning Model may be a systematic approach, it does not restrict a learner from expressing his creativity. I think what usually limits or encourages students in their expression of creativity are the activities which teachers use during their lessons. From what we have learnt in our lectures of Hybrid Learning Model previously, we have all seen that many different types of activities could be implemented in different phases. Furthermore, as Daniel has said, the activities which many of us have presented during our last tutorial also required the students to be creative and to think. I think that the Hybrid Learning Model is just a guide for teachers to use to plan their lessons in order to help their learners to understand and apply the concepts taught. The activities which we plan are just a means which help to fulfill the aims of the model. We can choose whether we want our lesson activities to encourage or limit creativity in our students. In fact, I think that in the use of any learning model, the promoting of creativity amongst students is highly dependent on the activities which the teachers planned.

Figure 10.7 Discussion board (systemic model restricts creativity?).

Thread: TSOI Hybrid Learning Model	Posted Date: January 23, 2010
Post: RE: TSOI Hybrid Learning Model	11:10 AM
Author: TAN SI YING JASMINE (BS20607)	Status: Published

Responding the concerns raised, for more effective use of the stoichiometry module, I feel that it can be carried out in one of the following ways: Teacher uses the module concurrently when conducting the lesson. Alternatively, the module can be broken down manually by the teacher, into smaller sub-sections within each section of the module (as what was mentioned) and the students can be asked to review the mini sections again at home.

Personally, I would prefer the module to be used concurrently with the lesson since it contains useful diagrams and problems, which the teacher can make use of to teach and guide the students. This would possibly minimize the occurrence of the problem of less motivated students not completing the module independently through e-learning. When it comes to calculation problems, students can be asked to work on the given questions individually, followed by think-pair-share and checking of answers with the teacher as a class. It is also possible that the teacher makes use of group learning to tone up the atmosphere of the lessons. The teacher can assign a few calculation problems to groups of 4 within the class and get the pupils to attempt after showing them a worked example, whenever necessary. Groups would be encouraged to share their solutions on the whiteboard and 2 points would be awarded to groups which have the correct solutions and any member of the group should able to explain the rationale for the steps taken. The points for each group would be tallied at the end of the chapter and the top 3 groups would receive a small token from the teacher. Having such healthy competition with proper rules set, might help the less motivated students to pay attention and follow the lesson closely, as the lesson is not structured in such a drill-and-practice manner anymore.

If one chooses to adopt the alternative method of using this stoichiometry module as mentioned earlier, technical modifications such as inserting of functions to pause or replay the narration could be done to make it even more user-friendly. By making the above modifications, the less motivated students might find it less taxing for them to go over the module again at home after the lesson, since they now have a nice blend of classroom learning and e-learning.

Figure 10.8 Discussion board (TSOI Hybrid Learning Model).

Thread: Implications of Using TSOI Hybrid
Learning Model on Teaching & Learning Chemistry
Post: RE: Implications of Using TSOI Hybrid
Learning Model on Teaching & Learning Chemistry
Author: FARN HSING CHIEH (BS20607)

Posted Date: January 25, 2010
11:03 AM
Status: Published

Elo all!

I agree that the TSOI hybrid learning model is an ingenious framework that assists us greatly in our lesson planning indeed. It helps us structure our lesson into an organized and systematic way and in some sense keeps me on track because the previous idea of introduction, development and conclusion is rather vague at times.

In response to Si Hui's previous thread on whether a systematic approach can dull creativity, I too feel that it may not be so. To me, the TSOI model is a guide to structuring the lesson and the success of the lesson depends on the type of activities and medium of delivery one will use to fill up each of the TSOI segments. Personally, I feel that the Translating phase poses the greatest challenge because it is at this point where we have to crack our heads to think of meaningful and interesting activity to engage them. This is the part where creativity can be at its best and for us to try out new teaching strategies. Sometimes, it is imperative that we assess not just the students' factual understanding of the lesson but also obtain feedback about the lesson to bring about improvements and if we should keep a particular teaching method in our repertoire.

What I gain most from this TSOI model is perhaps the part on operationalizing where students have to internalize the concept. In the past, I have also applied this process of getting students to create linkages between critical notions and concepts without knowing its significance. Thus this process was usually short and not thoroughly explored enough. Now that I have a better insight on what it is all about, I would be more mindful and be in a more capable position to come up with more activities to firm up this section of internalization.

In addition, I feel that the tone of delivery is of essence. One has to be high on empathy and low on criticism when giving feedback to the students. This will perhaps create an ethos that is nurturing and will enable each student to perform better as they are less likely to be afraid to be wrong.

Figure 10.9   Discussion board (Implications of using TSOI Hybrid Learning Model on teaching and learning of chemistry).

Thread: All discussions
Post: RE: All discussions
Author: NUR HUDA BINTE ISMAIL (BS20607)

Posted Date: January 23, 2010
2:52 PM
Status: Published

Other than flexible, like Hui Kheng mentioned, I also think it's quite dynamic.

In my ACB, we are going through the 5E Model of Inquiry (Engage, Explore, Explain, Elaborate and Evaluate). I realised that TSOI Hybrid Learning Model and 5E Model of Inquiry are quite similar and can be weaved with each other.

For example, in Translating, it is the experience stage experiential learning cycle. I think, we can then use and event or questiong to engage the students to "translate" the initial ideas and concept. I think there's alot we can do to explore in using both model.

What do you guys think about it?

Thread: All discussions
Post: RE: All discussions
Author: SITI AISYAH BINTE MUHAMMAD ALI
(BS20607)

Posted Date: January 24, 2010
5:38 PM
Status: Published

Yes the hybrid learning model is systematic yet flexible. The systematic flow of the learning model help guide us teachers in arranging our teaching ideas while the flexibility gives us the liberty on how to approach the different concepts of the topic.

As I was observing the other group's presentation on thursday, subconciously, there was a few times that I was actually re-arranging the groups' teaching ideas in my head. I was moving the ideas around between the 4 stages. This made me realise how flexible the hybrid learning model is because re-arranging the 4 ideas will actually lead to different lesson plans and delivery. Teachers are not restricted to a specific manner and are thus able to cater the lessons to the needs of their students while being systematic at the same time. :)

Thread: All discussions
Post: RE: All discussions
Author: NUR HUDA BINTE ISMAIL (BS20607)

Posted Date: January 25, 2010
12:23 AM
Status: Published

Hi girls,

Regarding Iffah's qn abt the 5E, to quickly summarise the 5E are Engage, Explore, Explain, Elaborate and Evaluate. It is one of the approach for Inquiry based lesson. This is a good site to explain in detailed: http://faculty.mwsu.edu/west/maryann.coe/coe/inquire/inquiry.htm

Something just came into mind for me. I saw that most of us felt that more questions are needed to test students' understanding because for this topic, it's is quite heavy on calculation. This makes me ponder that other than the flexibility and dynamism that we've mentioned, the TSOI Hybrid Learning Model, I do think, allows us to spend at one segment more than the other. It is not about strict allocation for each aspect. One example would be, for this topic, emphasis will be on Intergrating phase, because application would be the most important to check on students' understanding. One way, I would suggest is exposure to more questions because students will be able to apply the concept into calculation.

Figure 10.10 Discussion board (all discussions).

Blended Learning Design for Active Learning (Operationalizing and Integrating Phase)

In the third phase, Operationalizing phase, the group members collaborate actively in a 3-hour session, applying the principles of the TSOI HLM to provide the design of a chemistry lesson for secondary school level in the form of teaching ideas. The tutor has the essential task of not only facilitating the collaborative process to be active but also posing thought-provoking questions that are relevant to the group's ideas to elicit higher-order thinking. At the end of the collaborative process, the group members present their teaching ideas to the class for peer as well as tutor feedback and comments. These classroom experiences are for a concept-internalization process whereby all the critical features of the concept that is the hybrid learning model are linked together. This is essential so that the learner observes and experiences the meaningful functionality of the hybrid learning model. This is also to assist the learner in being operationally ready for further applications in the Integrating phase.

In the fourth phase, Integrating phase, Web 2.0, for example the Blog tool, is used for the purpose of reflective learning. The individual learner is to do a personal reflection based on one's solution concentration multimedia module's observations, asynchronous learning using online discussion boards, as well as the class activities on the hybrid learning model. A set of guidelines as to how one should frame the personal reflection to lead to reflective learning is provided and explained. Essentially, the learner is to reflect one's own thoughts on the hybrid learning model as an innovative pedagogy for teaching and learning in chemistry education. A week is given for this active learning activity. The Integrating phase places an emphasis on concept application for meaningful transfer of knowledge.

DISCUSSIONS

The critical features of the concept to be understood need to be first identified so that appropriate and meaningful activities can then be designed to lead an acquisition of the concept. With this in mind, the Translating phase of the TSOI HLM is a significant phase, as it presents the learner an initial preliminary awareness of the concept, which in this case is a hybrid learning model for teaching and learning to be learned. Yet equally, this phase poses the most challenge. This is so the instructional activities to be designed and experienced by the learner should be familiar to the learner so that one can make connections to one's existing knowledge structures. This would mean providing the appropriate experiences, whether it is face-to-face interactions or the in form of an e-learning module embedded in a learning content management system to be translated by the learner to a beginning idea of the concept. The learner's preliminary experience is then given more meaning in the Sculpting phase.

In the Sculpting phase, the beginning concept, experienced still in its un-refined form, is logically shaped or refined to a more concrete form by a se-ries of relevant and suitable instructional learning activities that are "crafted" meaningfully to assist the learner to identify the critical features of the con-cept to be understood. These instructional learning activities are also designed to encourage the learner to be actively involved in the appropriate thinking processes; for example, comparing and identifying patterns, abstracting, syn-thesizing, observing, and predicting, which the learner needs to complete to determine the critical features of the concept. As such, it is meaningful and relevant to enhance asynchronous learning by leveraging an online discussion board for its reflective nature so that learners have the time and opportu-nity to think through the contents of the discussion before expressing their thoughts, views, questions, and comments. Indeed, embedding the online dis-cussion board in the Sculpting phase and linking to the Translating phase as well as explicitly the Integrating phase, will enhance the quality of asynchro-nous learning, as illustrated in Figures 10.6 to 10.10. The thinking processes, be they critical thinking or reflective thinking, have evolved meaningfully dur-ing the groups' asynchronous learning using the online discussion board.

The Operationalizing phase serves to connect the Sculpting phase and the Integrating phase predominantly in the aspect of concept internaliza-tion. Indeed, there is a crucial need for the concept that is already con-structed to be internalized for meaningful functionality. Besides, an aware-ness of the problem-solving processes is also established within the group active learning via collaboration on the design of lessons to consolidate the functions and uses of the hybrid learning model. During the fourth phase, the Integrating phase, the just-learned concept, which is the hybrid learning model already internalized, is then put through the active process of reflective learning and is also integrated in different contexts in order for meaningful learning to occur. As the concept learned is a pedagogical model, it is applicable and meaningful to use Web 2.0, such as the Blog tool, in the reflective learning process.

Indeed, excerpts from the asynchronous learning responses lend sup-port to the potential of the hybrid learning model for enhancing asynchro-nous learning in a blended learning environment. Besides, studies by Tsoi (2007, 2008) have found a statistically significant difference between pre-test and posttest achievement means at the .05 level as they pertain to a learner's level of conceptual understanding of mole concept for each of the four groups using a multimedia learning package for learning of the mole concept, which has as its pedagogic model the TSOI Hybrid Learn-ing Model. It is likely that the four phases of the hybrid learning model together as a whole entity also have a positive overall effect on the conceptual learning of mole concept. In this preliminary study, the author recognizes the limitations that lie partly in the research rigor of the process, for example, empirical evidence and validation. However, this offers an alternative way of approaching the practice of enhancing asynchronous learning in a blended learning environment. The learner will build on the various concrete experiences and will learn how to create knowledge and integrate the knowledge with existing ideas and concepts in other contexts, and more importantly, be an active learner engaged in the various learning processes, including collaborative learning and reflective learning. Fundamentally, the TSOI Hybrid Learning Model has the functional potential capacity to give the educator an alternative practice model for enhancing asynchronous learning in a blended learning environment.

REFERENCES

His, S., & Agogino, A. M. (1994). The impact and instructional benefit of using multimedia case studies to teach engineering design. *Journal of Educational Multimedia and Hypermedia, 3,* 351–376.

Karplus, R. (1977). *Teaching and the development of reasoning.* Berkeley: University of California Press.

Kolb, D. (1984). *Experiential learning: Experience as the source of learning and development.* Englewood Cliffs, NJ: Prentice Hall.

Lawson, A. E. (1995). *Science teaching and the development of thinking.* Belmont, CA: Wadsworth.

Renner, J. W., & Marek, E. A. (1990). An educational theory base for science teaching. *Journal of Research in Science Teaching, 27*(3), 241–246.

Smith, D. M., & Kolb, D. A. (1986). *User's guide for learning-style inventory: A manual for teachers and trainers.* Boston, MA: McBer.

Tsoi, M. F. (2007). *Development and effects of multimedia design on learning of Mole Concept.* Doctoral thesis, Nanyang Technological University, Singapore.

Tsoi, M. F. (2008). Designing for engaged e-learning: TSOI Hybrid Learning Model. *The International Journal of Learning, 15*(6), 225–232.

Tsoi, M. F. (2009). Applying TSOI Hybrid Learning Model to enhance blended learning experience in science education. *Interactive Technology and Smart Education, 6,* 223–233.

Tsoi, M. F., & Goh, N. K. (1999). Practical multimedia design for chemical education. In G. Cumming et al. (Vol Eds.), *New human abilities for the networked society* (pp. 946–949). The Netherlands: IOS Press.

Tsoi, M. F., Goh, N. K., & Chia, L. S. (1998). Some suggestions for the teaching of the Mole Concept. In Wass, Margit (Eds.), *Enhancing learning: Challenge of integrating thinking and information technology into the curriculum* (pp. 778–785). Singapore: Educational Research Association.

Van Aalst, J. W., Carey, T. T., & McKerlie, D. L. (1995). *Design space analysis as training wheels in a framework for learning user interface design.* Proceedings CHI. New York: ACM, 154–162.

CPSIA information can be obtained at www.ICGtesting.com
Printed in the USA
BVOW010851010612

291560BV00002B/2/P